Andrea Maria Francesco Valli

The Three Domains of Life

A Brief History of Biodiversity

Andrea Maria Francesco Valli
Yzeure, Allier, France

ISSN 2731-8982 ISSN 2731-8990 (electronic)
Copernicus Books
ISBN 978-3-032-14801-8 ISBN 978-3-032-14802-5 (eBook)
https://doi.org/10.1007/978-3-032-14802-5

The original submitted manuscript has been translated into English. The translation was done using artificial intelligence. A subsequent revision was performed by the author(s) to further refine the work and to ensure that the translation is appropriate concerning content and scientific correctness. It may, however, read stylistically different from a conventional translation.

This Springer imprint is published by the registered company Springer Nature Switzerland AG
The registered company address is: Gewerbestrasse 11, 6330 Cham, Switzerland

Copernicus Books

Sparking Curiosity and Explaining the World

Drawing inspiration from their Renaissance namesake, Copernicus books revolve around scientific curiosity and discovery. Authored by experts from around the world, our books strive to break down barriers and make scientific knowledge more accessible to the public, tackling modern concepts and technologies in a nontechnical and engaging way. Copernicus books are always written with the lay reader in mind, offering introductory forays into different fields to show how the world of science is transforming our daily lives. From astronomy to medicine, business to biology, you will find herein an enriching collection of literature that answers your questions and inspires you to ask even more.

Dedicated to the memory of the Professors
Mario Ageno and Claude Guérin.
The first introduced me to biophysics,
the second taught me palaeontology

Foreword

The media often speaks the "Sixth Extinction of Species." This means that there were five more before that, as Monsieur de la Palisse would say... What were these five great mass extinctions? When did they occur? What caused them, long before humans existed? Which animal and plant species were involved? What were life and ecosystems like before, and how did all this begin?

These are just a few of the more than a hundred questions addressed in detail by this erudite scholar, passionate about the History of Life: Andrea Valli.

For him, the story of Life is far from a calm, linear river. Over the four and a half billion years of Earth's history, biodiversity is punctuated by crises whose outcomes were always unpredictable. The most severe occurred at the end of the Permian (−252.6 Ma), the final phase of the Primary Era, while the last led to the extinction of the dinosaurs, at the end of the Secondary Era (−65 Ma). These crises, however, are precisely what define the eras as delineated by palaeontologists.

A Long History

These crises punctuate the long and surprising evolution history of living species, carefully described by the author. After a lengthy biochemical phase lasting two billion years (with the precision of a few million years), life slowly organized into structures ("cells") lacking a nucleus (what we now call "bacteria"). Later, the first eukaryotic cells (with nucleus) appeared, gradually giving rise to became multicellular organisms.

About 540 million years ago, life underwent a "firework" of diversification, leading to the emergence of all the major groups we recognize today. Until then confined to water, life finally invades the land. Later (200 million later!) Insects and Vertebrates conquered the skies. Finally, humans appeared very late on the chronological scale, only three million years ago.

During this entire period, 99% of the species that have existed on Earth became extinct. We know this through the fossils preserved in the Earth's strata. Each extinction was followed by the emergence of new, increasingly diverse life forms. Mass extinction, therefore, played a decisive role in the diversification of living beings, as the author meticulously demonstrates.

However, the History of Life proposed by Andrea Valli is not merely descriptive or linear. Scientific to the tip of his fingernails, the author avoids the sensationalism of palaeontologists who prioritize headlines over rigorous research. He advances hypotheses cautiously, weighing their pros and cons, and critically examines every result, detailing its limits and scope.

With pedagogical care drawn from his teaching experience, Andrea Valli illustrates complex ideas with figures or schemes. In addition, each chapter includes a summary, guiding the reader through this vast and intricate subject.

Revelations

Valli's History of Life offers many insights, even for experienced readers. For example, defining life is not straightforward. The standard definition (that living things feed and reproduce) is incomplete. Valli references a more comprehensive definition proposed by the Italian scientist Ageno, little cited in France, and discusses it at length. He raises fundamental questions: When does life begin? How did it arise? How did molecules transition into proto-organisms?

Two hypotheses, inspired by Ageno, suggest that "active precursors for peptide chain synthesis and the establishment of a code linking amino acids to nucleotide triplets" may have led to life. In this scenario, life emerges as a chain of events, each enabling the next, given favourable initial conditions (whether on Earth or elsewhere). While the hypothesis of life originating elsewhere remains plausible, it merely shifts the question: What happened there? And can we find traces of it?

The first organisms are also discussed, though their fossils are scarce due to their fragility and microscopic size. The concept of LUCA, the Last Universal Common Ancestor, is introduced. LUCA existed between 3.8 and 3.5 billion years ago, prior to the divergence of bacteria, archaea[1] and finally eukaryotes, to which we belong.

While its exact form remains unknown, current research aims to reconstruct it from the genomes of its descendants.

According to the most accredited scenario by the scientific community, bacteria would have been the first to diverge, while another branch would have led first to archaea and then to eukaryotes. Was LUCA's genome made up of RNA or DNA? All this remains speculative, because the traces collected reveal nothing about it. However, as the author tells us, "These hypotheses have the merit of constituting excellent foundations for future research on the subject." As for the elusive LUCA, he has certainly disappeared, but some of his features certainly survive in the genome of his distant descendants. Current researchers are therefore trying to reconstruct it by climbing the tree of the living.

The author also addresses viruses: Are they living? The question remains open, but their role in the genomes of bacteria, archaea, and even humans is significant. Approximately, 13% of bacterial and archaeal genomes derive from viral sequences, rising to 40% in humans. Could we consider ourselves "a species of GMO shaped by evolution"?

Starting from the end of the Archaean, while bacteria and archaea continue their more or less discrete evolution, new entities emerge. Eukaryotes (organisms with nuclei and organelles) likely arose through symbiotic associations of prokaryotic cells. Some researchers remain sceptical of this theory, yet eukaryotes occupied niches inaccessible to prokaryotes, enabled by increased size and specialized tissues.

The evolution of animals, plants, and fungi continues inexorably! Eukaryotes and prokaryotes survived climate changes, volcanic eruptions, meteorite impacts, and early atmospheric transformations, such as the oxygenation of the Earth, a metabolic byproduct initially toxic to life but later integrated into energy production, facilitating the proliferation of life.

Millions of years of diversification ensued during the Primary Age, with some groups surviving to the present and others succumbing to extinction. Andrea Valli introduces them meticulously, maintaining a critical scientific perspective.

---

[1] Bacteria and archaea were once united in a single group. They are all prokaryotic beings; their cell, that is, has no nucleus.

Of course, it could not be concluded without the History of Primates and the History of Man. Not only *Homo sapiens*! For the author, human evolution is not a linear progression from ancestor to modern forms but a series of episodes with multiple contemporary hominid species, sharing environments the history of our evolutionary line.

He concludes, insightfully, with a reflection that we endorse: "There is no rational logic in the evolution of living beings." The most abundant organisms in ecosystems are the oldest groups, particularly bacteria, whose survival and adaptation outlasted many more complex forms. The succession of being more complex has not made them disappear. Not at all: on the contrary, they have continued to develop, adapt, and even thrive together with the newcomers!

Finally, based on current knowledge, the author cautiously asserts that we cannot predict the long-term evolution of biodiversity. Yet, history suggests that life will undoubtedly persist and diversify. Should a sixth mass extinction occur, new life forms will eventually emerge, potentially more attuned to coexistence with nature, leaving traces for future observers to learn from our mistakes!

André Giordan (1946–2023)
Founder of the Laboratoire d'Épistémologie et de Didactique des Sciences and Senior Chairman of the IUBS CBE (International Union of Biological Sciences)
Professor at the University of Geneva
Geneva, Switzerland

Acknowledgements The genesis of this book stems from a collaboration with Prof. André Giordan, of the Faculty of *Epistémologie et de Didactique des Sciences* of the University of Geneva, during my work for the *Conseil départemental de l'Allier* (in the department of the same name). I was employed by this Institution between 2003 and 2007, to carry out an ambitious project: the creation of a scientific park dedicated to the history of Life on Earth, from its origins to the present day. The evolution of terrestrial biodiversity was the central axis of the program. Unfortunately, the project was not realized in its initial conception. Between 2009 and 2010, the topic regained attention. During this period, a much more modest project was conceived and, due to my experience in the field, I was asked not only to oversee the drafting of the scientific program but also to contribute to the realization of a temporary exhibition for 2013. However, for various reasons, the scientific project was significantly "scaled down" and I no longer recognize it in my own. Therefore, the idea of creating a text capable of presenting the concepts initially envisaged for the park is back resurfaced.

The creation of this book is primally due to A. Giordan, who encouraged me to put in writing what was originally conceived for the project. He supported me during the drafting of the manuscript, reading and correcting it several times, trusting in my abilities as an author and populariser, and ultimately agreeing to write the preface of the work. Without him, this book would never have existed.

I would also like to thank *Dédale Editions* and, in particular, François Ové, who, with rigor and professionalism, guided me through the composition of the manuscript and published the volume.

Given the vast scope of this work, external help was essential. Each year, a significant amount of new data is added to the history of our planet, and particularly to the Life it harbours. I had to collect extensive bibliographic material. In this regard, I am especially grateful to those who, voluntarily or involuntarily (since the accumulation of data began long before the writing of the text), provided the material necessary for drafting the manuscript. Among them are André Nel and Jean-Sébastien Steyer of the *Muséum national d'Histoire naturelle* of Paris; Claude Guérin, Bertrand Lefebvre and Abel Prieur of the *Université Claude-Bernard Lyon 1*; Bruno Corbara of the *Université Blaise Pascal* of Clermont-Ferrand; Philippe Fernandez of the *Maison méditerranéenne des Sciences de l'Homme* of Aix-en-Provence; Patrick Forterre of the *Institut Pasteur* and of the *Université Paris Sud*; Pierre Mazeyrat of the *Société Scientifique du Bourbonnais*; Maria Rita Palombo of the *Università di Roma "La Sapienza"*; Kenny J. Travouillon of the University del

Queensland, Australia; George O. Poinar Jr. of the Oregon State University; J. William Schopf, of the University of California; Linda A. Amaral-Zettler of the Marine Biological Laboratory, Massachusetts; Eric Delson, of the American Museum of Naturel History and of the City University of New York. These researchers generously responded to my requests for information and access to publications. Without their help, I could never have obtained all the necessary information for this work. I apologize if I have unintentionally omitted any names.

I also wish to thank my father, Giulio Valli, who greatly assisted me in locating much of the bibliography cited here, particularly articles published in *Nature* and *Science*. He also sent me unsolicited texts, which proved highly relevant.

Finally, I am grateful to my friends Marc and Mireille Breton, Françoise Aimard-Charcot, Jacques Aimard, Roland Lafont, and Monique Châtelet, for their patience in reading and commenting on my manuscript. Their judgment and suggestions were invaluable. If this text has become accessible to a broad audience, it is largely thanks to their feedback.

The English edition owes much to my dear friend Eugenio Mieli, who helped introduce the book to Springer and facilitated the translation using Claude (a virtual assistant created by Anthropic), to my colleague (and friend) Lionel Lajerige, who kindly reviewed the entire English text, and to Marina Forlizzi, Barabara Amorese, Antony Raj Joseph and the whole editorial team.

I am deeply grateful to all those mentioned. Of course, I remain solely responsible for any errors and inaccuracies that may still be present in this volume.

Competing Interests The author has no competing interests to declare that are relevant to the content of this manuscript.

Introduction

What is the purpose of studying the history of Life? Why should we take an interest in events that occurred long ago and are often difficult to interpret? Knowledge of the past is essential for understanding the present and for answering fundamental questions such as: "Who are we?" "Where do we come from?" Of course, all this does not concern only the history of a single country (such as France) or of our own species (the history of humankind), one species among many.

There is a close interconnection between living beings and the physical phenomena that occur on our planet. Therefore, in order to decipher what happened in the past, interdisciplinary studies (concerning biology, geology, chemical analyses) are required. Only in this way is it possible to reconstruct events (such as climate evolution) that in turn make it possible to propose models and make predictions. All these approaches are necessary to evaluate "Biodiversity," whether present today or in remote times.

The concept of biodiversity is very much in the spotlight today. In fact, many voices warn of the dangers that threaten it. But has our planet not already faced similar situations? To answer this question, we must look to the past and ask what happened in more or less distant times.

Let us begin, however, by asking what biodiversity is? What does this term actually mean? The following definition was proposed during the XVIIIth General Assembly of the IUCN, "The World Conservation Union," held in Costa Rica in 1988. "Biological diversity, or biodiversity, refers to the variety and variability of all living organisms. It includes the genetic variability within each species and its populations, the variability of species and their life forms, the diversity of associated species complexes and their interactions, and finally the ecological processes that influence these complexes or in which they play a role (ecosystem diversity)".[1]

All this means that, in order to assess the biodiversity of a given geographical region, it is necessary to consider all the living beings present, since each one is different from the others (even if only slightly). Each individual contributes to overall biodiversity through its unique features contribution. In essence, biodiversity is proportional to the total living mass of a giving region.

However, this is not the only definition in use. Evaluating all the organisms present in a given community or geographical area is not always easy: sometimes it is impossible (consider, for example, ecosystems from past eras that have literally disappeared).

Another approach is to consider the number of species present (specific biodiversity): the more species, the richer the biodiversity.

Yet biodiversity cannot be reduced simply to a count of individuals or species present in a certain locality; all this is simply reductive! To fully appreciate it, we must also consider the interactions among living beings, whether of the same or of different species, together with the changes driven by transformations in their environments.[2] This will be the red thread that I will follow in my narrative.

In any case, whatever definition we adopt, biodiversity always concerns living organisms. But do we really know them well enough? Are we able to define these entities precisely? This book aims to recount the history of the evolution of living beings, which is also the history of biodiversity: the stages it has gone through and the challenges it has faced.

This is a way of understanding how today's ecosystems and their interactions came to be. Since the history of Life is a vast subject, and not all its stages have been fully clarified (not to mention the limits of my own knowledge), this account will eventually remain partial.

---

[1] I am grateful to Prof. André Giordan for passing on this definition. It came to me during the joint work for the *Conseil général de l'Allier*. A few years later, during a lecture by Prof. Bruno Corbara of the Université Blaise-Pascal in Clermont-Ferrand, held for the event *Journées Nature d'Avermes*, in Allier (France), I discovered that, in fact, there are different definitions of biodiversity.

[2] This concept is masterfully set out in the book by Robert Barbault (2006), a very informative reading for all those who wish to deepen the subject. Another classic text dealing with biodiversity is that of Edward O. Wilson (2001). Finally, for all those interested in the interactions between different organisms, I recommend the excellent volume of Prof. Marc-André Selosse (2017).

A lifetime would not be long enough to cover everything known in this field. Moreover, it is a constantly evolving topic: every year new discoveries enrich our overall viewpoint.

Of course, there is no shortage of works on the history of life, written in various languages. Why, then, publish another? My academic career and personal experiences have given me a perspective somewhat different from that of other authors. In addition, in this book, I wish to present the hypotheses put forward by Prof. Mario Ageno during the 1990s, which received little attention, not only abroad (for lack of translations), but even in Italy.

Yet these ideas deserve to be reconsidered or, at the very least, known by a wider audience. Finally, teaching a subject or writing a book about it is undoubtedly one of the surest ways to test one's own knowledge of it...

The purpose of this book is to awaken readers' interest in these questions and to provide references and guidelines for those who wish to explore them further. I hope it will be stimulating even for non-specialists (provided they have some basic scientific knowledge and, above all, curiosity about living beings).

In the following pages, I will try to use language as simple as possible. However, the subjects themselves are complex, and I must warn you: if you wish to follow, you will need to remain attentive. To avoid making the reading too heavy, more detailed discussions will be relegated to footnotes (which can be consulted according to the rhythm of reading) or to appendices at the end of each chapter. I have also indicated some entries from the informatic encyclopaedia "Wikipedia," in which people can deepen certain information. Please note, however, that I consulted these items in 2025 and now they could be subject to change.

Finally, I have added a summary at the end of each chapter to help recap the main interesting points.

References[3]

*Barbault, R. 2006. *Un éléphant dans un jeu de quilles - L'homme dans la biodiversité Seuil*, Points Sciences, Paris.

*Selosse, M.A. 2017. *Jamais seul - Ces microbes qui construisent les plantes, les animaux et les civilisations*. Actes Sud, Arles.

*Wilson, E.O. 2001. *The Diversity of Life*. 2nd ed, Penguin Books, London.

[3] Works preceded by an asterisk can be read, more or less easily, by people with a non professional skilling.

Contents

1

The Transition from the Non-living to the Living

1.1 Defining a Living Being

Without any doubt, each of us can distinguish a living being from an inanimate object. A lion, a fly, a fungus, or even a bacterium would naturally be placed in the first category. On the other hand, objects such as stones, fire, or water are clearly regarded as non-living. Moreover, while a tree as a whole is a living organism, a broken branch is nothing more than an object. A single part of an animal or plant cannot be considered a living being, but only a fragment of one, having lost the properties that once made it alive.[1] The same is for a dead animal: the moment death occurs, it loses all the qualities that characterized it as living.

But what exactly are these capabilities? How can we answer our students, our children, our friends when they ask for clarification? Let us try to give a clear answer.

In the past, a living being was often defined by its ability to feed and reproduce. Every organism (plant, animal or microorganism) needs nourishment and has the ability to reproduce; otherwise, the species would vanish.

[1] In fact, some living beings possess extraordinary regenerative capacities: for example, thanks to autotomy (the possibility of voluntarily losing a part of their body and then regenerating it) certain organisms can "reproduce". Take a starfish that mutilates itself with one arm. Not only can the star regenerate its appendix, but the lost arm can in turn produce the whole individual. This particular "breeding method" is not limited to echinoderms. For example, it exists also among earthworms. There are cases, therefore, where even a simple part of a living being retains its particular abilities and can completely regenerate an entire organism, perfectly viable.

A. M. F. Valli, *The Three Domains of Life*, Copernicus Books,
https://doi.org/10.1007/978-3-032-14802-5_1

Yet this definition is not sufficient. For example, fire seems to "feed" and "reproduce". In fact, a flame consumes wood or another type of fuel to stay alight. If we place a torch or a piece of paper next to it, the fire spreads, creating new flames. And so below… This is like reproduction, since the original flame gives rise to others.

And yet, nobody would seriously classify fire as a living being. Why not? Because its "feeding" and "reproduction" differ fundamentally from those of plants, animals, or microbes. Clearly, something is missing.

The Italian physicist Mario Ageno addressed this issue in the 1980s and 1990s[2] by proposing a more precise definition. That allows us to characterize all living organisms from the Earth, fossil and recent. And only them.

According to Ageno, a living being is, above all, a physical system with clearly boundary (its body, whether as large as a tree trunk or as small as a bacterial cell). Within these boundaries, food, gases, and heat are taken in, while waste and excess heat are expelled. Inside, a highly ordered set of chemical processes occurs, in space and time (the author speaks of the "coherence" of internal processes). This coherence is essential: without it, metabolism would collapse into chaos, leading to death.

What ensures this order? Every living organism possesses a "program", that encodes its components and regulates its metabolic processes. In modern terms, this program is DNA.[3] All living organisms possess DNA, which not only preserves their genetic heritage but also regulates development and metabolism (deciding what to produce, when and how), in response to internal or external signals.

Directly or indirectly, DNA plays a central role in controlling metabolic processes. This function must not be forgotten or diminished in relation to the task of preserving genetic information. Without the order due to DNA, a living organism would soon cease to be such!

Thus, to borrow Ageno's words, a living being is: "an open physical system (capable of exchanging materials and energy with its environment), hosting coherent chemical processes and equipped with a program".

---

[2] Mario Ageno's definition of a 'living being' is presented in three fundamental works (Ageno 1986, 1991, 1992a). The interested reader can read it with interest.

[3] This does not mean that the DNA of an organism should be reduced to a simple "instruction manual", like a trivial algorithm. DNA possesses extraordinary properties capable of conferring on it an almost unequalled plasticity, inside a cell. However, it cannot be denied that among its most important properties is also the ability to function as a program. In fact, DNA regulates the "life" of cells: it determines, aided by external stimuli, which substances have to be produced and when the process should begin. Perhaps a more picturesque (but also more consistent with reality) definition would be to assimilate the DNA to an orchestra conductor, who knows the partitions and directs the set of instruments that make up this extraordinary machine which is the living being.

This program is hereditary. In asexual reproduction (e.g., cell division) the genetic material is transmitted in full; in sexual reproduction, roughly half is inherited from each parent.[4] Errors sometimes occur, producing mutations, while recombination generates unique combinations of genes. Every living being therefore has a unique program of its own[5] (this is true for all sexually reproducing organisms; for those that multiply by cell division it is only possible). Those better suited to their environment reproduce more successfully leaving more descendants (the foundation of biological evolution[6]).

Importantly, ALL known living organisms, from bacteria to dinosaurs, satisfy these criteria, whereas inanimate phenomena do not.

For example, fire has no defined body and no internal program. A river flows in an orderly way, inside its banks, but it is not the site of chemical processes regulated from within. A computer, on the other hand, has the program, but it carries out mechanical and electrical processes, not biochemical ones. To be alive, all criteria must be met together, not just some. Otherwise, we return to the case of fire, which, in its own way, is capable of "feeding" and "reproducing", but which is not a living being.

To end this section dedicated to the characteristics of living beings, I want to point out that this is ONE definition, not THE definition. In fact, it

---

[4] However, there are exceptions for sexually reproducing organisms. We find it, for example, in the case of haplodiploid organisms (ants, bees, wasps and other insects). Take the example of ants and bees. The characteristic of these insects is that they have female individuals with genetic heritage consisting of a half of maternal origin and the other half of paternal origin (exactly like all other diploid living beings, that is those who have two copies of their genome). Male individuals, however, only have the maternal copy. To produce males, bees and ants do not need to mate. Instead, they must be fertilized to produce female individuals. It is clear that, for ants and bees, the general rule must be changed in order to be applied successfully (Wilson 1975). I will return to sexuality and cell division in Chap. 3.

[5] Consider a population of sexually reproducing organisms whose genome (the program) is composed of 5000 genes (certainly not an excessive number, since humans possess about 30,000) and establish that each gene may have two alleles (an allele is a particular variation of the gene in question). Possible combinations for the resulting genomes are 2^{5000} (this value only takes into account possible genetic combinations, not to mention the random mutations which may increase the number even further). If we express this value with the powers of 10, we get ~ 10^{1500}. This is an astronomical number: millions of times larger than all the nucleons in the Universe. Nucleons are the elements that make up atomic nuclei (protons and neutrons). The maximum number of atoms that can form at the same time in our Universe is necessarily lower than the number of nucleons. It can therefore be excluded that, within the same population, there are two organisms with the same genetic heritage (an exception is still the identical twins). Not only that, but it is also unthinkable that a particular combination has happened more than once in the course of the history of a species.

[6] In this work, I do not intend to deal with the theory of biological evolution through natural selection developed by Charles Darwin, nor to illustrate its recent developments. This goes beyond the aims of this book. However, those interested in studying the subject or simply to deepen it can consult with profit, not only the work of Darwin (2009) but also the volume published under the direction of Lecointre (2009). For those readers who have ease with English I also recommend the book by Jablonka and Lamb (2006).

is possible to formulate others as well.[7] The advantage of the definition of Ageno is that it allows to easily recognize all known organisms, extinct or living, and to exclude inanimate objects or considered as such.

Finally, the basic unit of Life is the cell, equipped with a biological apparatus. Anything not made of cells cannot be considered alive. For this reason, "viruses"[8] (although made of genetic material enclosed in a protein shell—the capsid) are excluded under the classical definition, since their lack the cellular machinery needed for metabolism. They can only reproduce by hijacking the machinery of a host cell, where they can multiply their DNA molecule (or RNA, in the case of "viruses" which, like AIDS, store their generic information in an RNA molecule) and capsid components. Once this is done, the host cell is destroyed and a multitude of new "viruses" emerge to infect new targets.[9]

1.2 The Transition from Non-living to Living: Biological Molecules

Louis Pasteur and later scientists demonstrated that every living being comes from another living being, whether sexual reproduction or cell, Life arises from pre-existing Life. The old belief in spontaneous generation (the idea that Life could emerge directly from inanimate matter) was proven false.

But then, where did Life come from in the first place? Has it always existed on Earth, or in the galaxy? In reality, we have to look at things differently. Let us go back to the observation expressed at the beginning of the paragraph, namely that every organism derives from another living being. Evidently, it is the current conditions that reign on our planet that prevent the transition from the non-living to the living. If these were changed, we should be able to reverse the trend and make possible what is not.

However, this process is long and complicated, and it needs many steps. And, like all phenomena, it must be subject to the laws of randomness

[7] Other formulations of Life are presented in an article by Forterre (2010). However, most of them consider a living entity with a functional cell apparatus. This is almost equivalent to the definition of Ageno.

[8] In this chapter, the term "virus", as defined classically, will be indicated in quotation marks. However, the definition of these entities has evolved and a different interpretation has recently been proposed. Viruses, according to the new definition, would be fully-fledged living beings. I will revisit them in paragraph 2.4.

[9] I will return to the "viruses" in paragraph 2.4 to give you a new interpretation of these entities. For the moment, it is enough to know that the definition of living being given excludes from the list their classical conception. This may seem arbitrary, but it has the merit of being clear and making unambiguous choices.

governing the physical world, of which biology is but a manifestation.[10] The problem must therefore be reformulated in a different way: is it possible to conceive conditions such as to allow, step by step, the transition from a situation characterized by the absence of living beings to one in which organisms are able to emerge spontaneously, without the need for pre-existing organisms? We think it possible. But then did these conditions really occur during the evolution of the physical world, on our planet, or in another part of the Universe? The answer is obviously positive, positive because we are here. So, the question must be reformulated: when and where, did these conditions occur?

Let us examine things in order. Living beings are chemical systems built from specifical molecules: the biological ones. They have been so called because they are the product of biological activity: only organisms would be able to synthesize them. These includes proteins, nucleic acids, lipids (fatty acids) and sugars. These "building blocks" are largely composed of just four elements: hydrogen (H), carbon (C), oxygen (O) and nitrogen (N), which together make up about 96% of living matter (Ageno 1991). These elements are part of the most abundant in the entire Universe! To these, we must add phosphorus (P) and sulphur (S) essential though less abundant (they account for only 0.9% and 0.3%, respectively). Other elements such as sodium (Na), potassium (K), magnesium (Mg), calcium (Ca), iron (Fe), and others also play vital roles, for example, in enzymatic functions or cellular signalling. In addition, each element may have a particular importance for a specific group of living organisms.

Among the molecules mentioned above, let us limit ourselves for now to proteins and nucleic acids. They are all macromolecules, whose mass exceeds by hundreds or thousands of times that of the hydrogen atom, the lightest element in the Universe.[11]

Proteins are chains made up of monomers, i.e. units linked like beads on a string. The result is a complete and functional product. The monomers making up proteins are amino acids, molecules at the centre of which there

---

[10] To appreciate the links between biology and other scientific disciplines (and, in particular, physics), the curious reader can read Ageno (1992b). In the work cited, the physicist presents a brilliant demonstration of the "independence" of biology from physical sciences.

[11] The hydrogen atom is generally made up of an electron (negatively charged particle) and a proton (positively charged particle). The mass of the electron being negligible compared to that of the proton, it is the weight of the latter which determines that of hydrogen. However, some hydrogen atoms may have one or more neutrons (electrically neutral particles). The weight of a neutron is equal to that of a proton. Thus, in order to estimate the weight of various atoms or molecules, it is sufficient to count the number of protons and neutrons that make up them. This will tell us how many times their mass exceeds that of the standard hydrogen atom (an electron plus a proton) which is considered as the unit (= 1).

is a carbon atom. Its four bonds are formed with a hydrogen atom (H), a carboxyl group (COOH), an amino group (NH_2), and a radical (R). The latter can vary and it is precisely its composition that characterizes each different amino acid (Fig. 1.1). There are 20 amino acids used by living organisms in all. Despite this limited set, the possible combinations are astronomical. A typical protein has around 300 amino acids, giving rise to about $20^{300} \approx 10^{390}$ possible sequences, though only a tiny fraction is functional.

Nucleic acids, by contrast, are built from just four different bases: adenine (A), guanine (G), cytosine (C) and thymine (C), in AND (Fig. 1.2). These bases are attached to a sugar with 5 carbon atoms (deoxyribose in DNA,

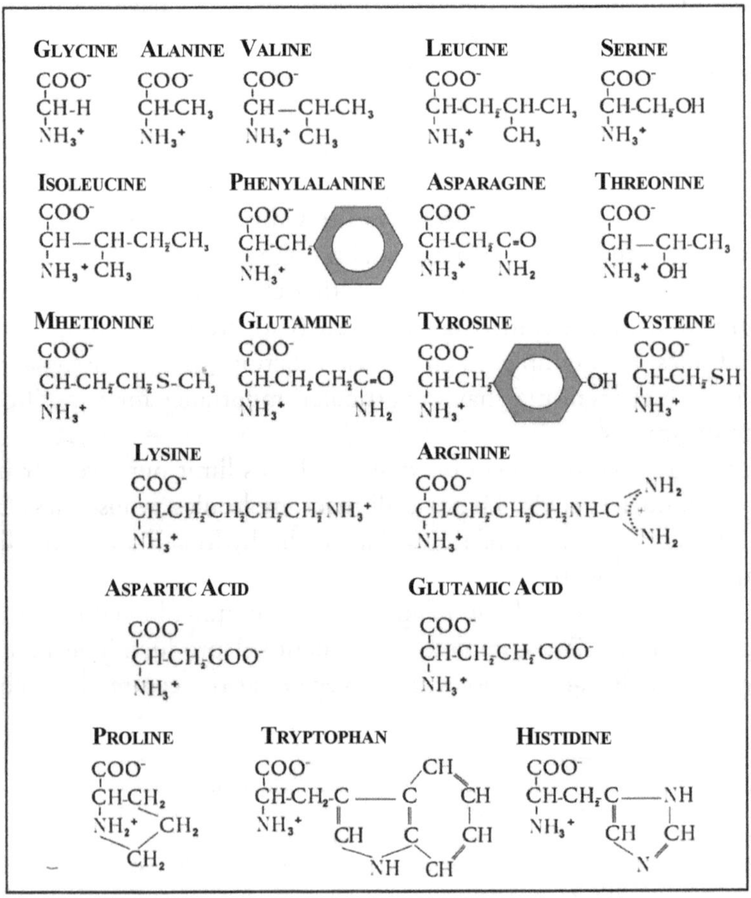

GLYCINE: COO^- / CH-H / NH_3^+

ALANINE: COO^- / $CH\text{-}CH_3$ / NH_3^+

VALINE: COO^- / $CH—CH\text{-}CH_3$ / NH_3^+ CH_3

LEUCINE: COO^- / $CH\text{-}CH_2\text{-}CH\text{-}CH_3$ / NH_3^+ CH_3

SERINE: COO^- / $CH\text{-}CH_2OH$ / NH_3^+

ISOLEUCINE: COO^- / $CH—CH\text{-}CH_2CH_3$ / NH_3^+ CH_3

PHENYLALANINE: COO^- / $CH\text{-}CH_2$ / NH_3^+

ASPARAGINE: COO^- / $CH\text{-}CH_2C\text{=}O$ / NH_3^+ NH_2

THREONINE: COO^- / $CH—CH\text{-}CH_3$ / NH_3^+ OH

MHETIONINE: COO^- / $CH\text{-}CH_2CH_2S\text{-}CH_3$ / NH_3^+

GLUTAMINE: COO^- / $CH\text{-}CH_2CH_2C\text{=}O$ / NH_3^+ NH_2

TYROSINE: COO^- / $CH\text{-}CH_2$ -OH / NH_3^+

CYSTEINE: COO^- / $CH\text{-}CH_2SH$ / NH_3^+

LYSINE: COO^- / $CH\text{-}CH_2CH_2CH_2CH_2NH_3^+$ / NH_3^+

ARGININE: COO^- / $CH\text{-}CH_2CH_2CH_2NH\text{-}C$ NH_2 + NH_2 / NH_3^+

ASPARTIC ACID: COO^- / $CH\text{-}CH_2COO^-$ / NH_3^+

GLUTAMIC ACID: COO^- / $CH\text{-}CH_2CH_2COO^-$ / NH_3^+

PROLINE: COO^- / $CH\text{-}CH_2$ / NH_2^+ CH_2 / CH_2

TRYPTOPHAN: COO^- / $CH\text{-}CH_2\text{-}C—C$ CH CH / NH_3^+ CH C CH / NH CH

HISTIDINE: COO^- / $CH\text{-}CH_2C—NH$ / NH_3^+ CH CH / N

Fig. 1.1 The 20 amino acids used by all living organisms. At the centre of each molecule is a carbon atom bonded to a hydrogen atom (indicated CH). The –COOH group is presented in its deprotonated form (–$COOH^-$) and the –NH_2 group is shown in its protonated form (–NH_3^+)

ribose in RNA) and linked by phosphate groups (here we can appreciate the importance of this element for the living world!). DNA's unique double helix arises from its deoxyribose backbone, a structure that ARN lacks. In reality, DNA and RNA also have another difference: thymine on the first molecule, uracil (U) on the second. But, in this case, it is really a detail.

An important aspect for the functioning of these molecules is that the bases are complementary: adenine pairs with thymine/uracil, while guanine with cytosine. This means that a DNA strand with the 'thymine-guanine-adenine-cytosine' can associate with another strand with the complementary sequence 'adenine-cytosine-thymine-guanine' (Fig. 1.3). This results in a double-stranded molecule that can fold in on itself in the form of a double helix (Fig. 1.4). Once again, only four different bases are sufficient to produce can produce all the genetic variation known in the world.

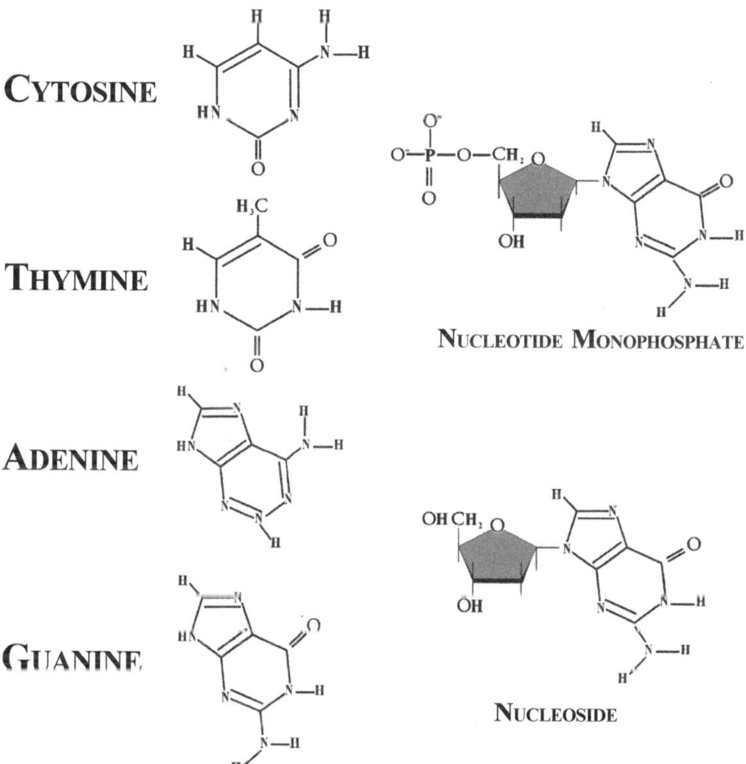

Fig. 1.2 The four bases of DNA are presented on the left. At the bottom right, a DNA nucleoside (the base, here guanine, plus the deoxyribose sugar). At the top right, the same nucleoside with an added phosphate group (= nucleotide monophosphate). Redrawn from Hebsgaard et al. (2005)

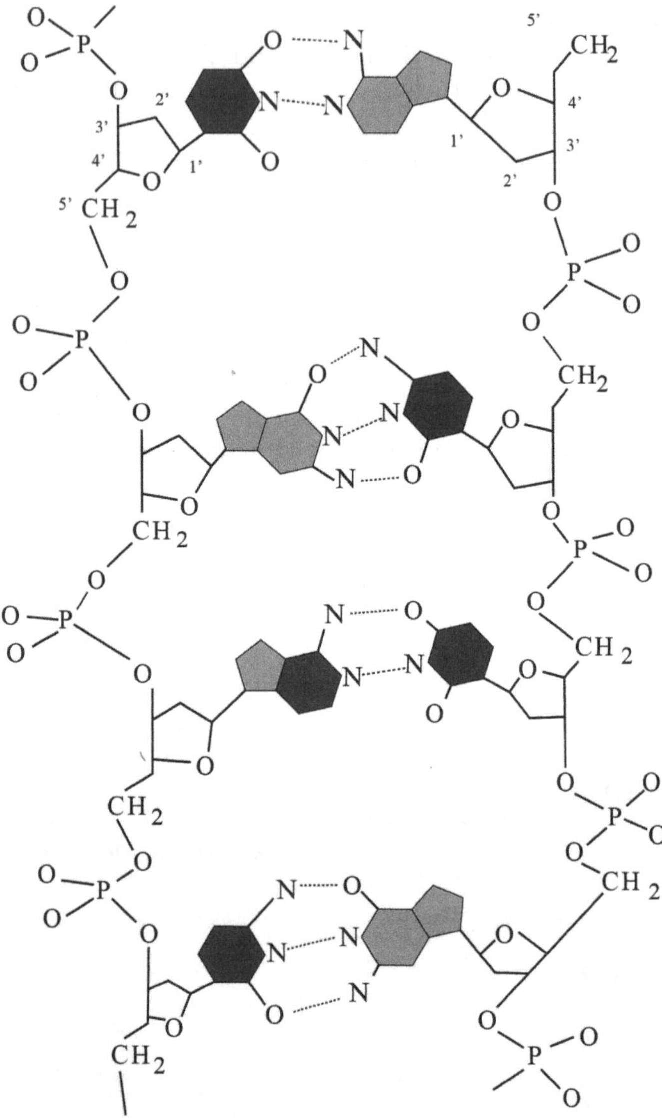

Fig. 1.3 Double-stranded DNA. The left strand shows, from top to bottom: thymine-guanine-adenine-cytosine. In the right strand shows the complementary sequence: adenine-cytosine-thymine-guanine. Redrawn starting from Geis (1983)

In a protein, we can identify the active site, which is the region where the molecule's enzymatic activity occurs (an enzyme is a biological catalyst, i.e., an agent that facilitates and accelerates a spontaneous chemical reaction) and the rest of the molecule, which forms the protein body, protecting and surrounding the enzyme centre. This particular structure allows the protein to

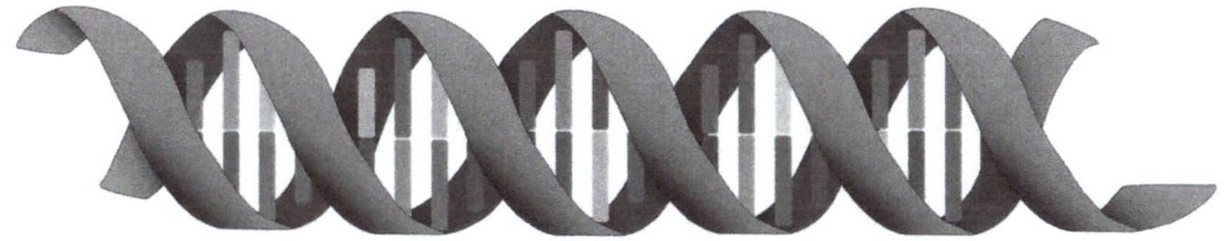

Fig. 1.4 The double-stranded DNA helix. The backbone (skeleton), in grey, is formed by sugar molecules and phosphate group. The vertical "steps" represent the four DNA bases, as shown in Fig. 1.2. Redrawn from Hebsgaard et al. (2005)

perform its activity optimally: to be effective while remaining highly specific, i.e., interacting with only a limited number of other molecules (often only one). The three-dimensional shape of a protein depends on the environment in which it was formed and the way it was synthesized. All these factors are crucial for the proper functioning of the molecule.

On the other hand, the DNA molecule, in aqueous solution, adopts a relatively stable double-helix structure, regardless of the base sequence. Through the genetic code (a combination of three nucleotides corresponds to an amino acid), ADN stores all the information necessary for protein production and also regulates the synthesis process.

It is precisely through the protein/amino acid interaction that Life is realized, according to the definition given earlier: if proteins perform chemical functions within on organism, DNA, in turn, controls coherence and functions as cellular "software".

After this brief introduction to biological molecules, let us consider the following question: is it possible to produce these molecules artificially, outside of a biological context? Since 1828, the answer has been yes! At this time, the German chemist Friedrich Wöhler succeeded in synthesizing urea in the laboratory. This was only the first step; today, an increasing number of biological molecules are produced in the laboratories through processes that have become routine.

Another important step was made by Stanley Miller, who, in 1952–53, prepared a gaseous mixture of water vapor (H_2O), carbon dioxide (CO_2), methane (CH_4) and ammonia (NH_3), intended to simulate the primitive atmosphere surrounding Earth during the early stages of the crust cooling. This mixture was subjected to electric discharges in a specialized experimental apparatus (Miller 1953). The result was a liquid containing several organic molecules. Subsequent repetitions of the experiment, with modifications to the parameters (especially the composition of the starting gaseous mixture) demonstrated that in an oxygen-free atmosphere (without the O_2 gas), by providing appropriate energy, it is possible to synthesize biological molecules

(Ageno 1991, paragraph 2.6; Fry 2000, Chap. 7). Of course, the quality and quantity of these molecules depend on the starting mixture; with the best results achieved with a reduced atmosphere, in which the main elements are bound to hydrogen.

Subsequent research has shown that, at the time relevant to our discussion, Earth likely had a different atmosphere from that used in Miller's experiments. The early atmosphere mainly consisted in CO_2, H_2O and N_2 (nitrogen gas). Under these conditions, artificial synthesis experiments like Miller's produce liquids with very few biological components.

However, during the same time (the solidification of the Earth's crust and its cooling, between 4.2 and 4 billion years), when the oceans were forming and the activity of underwater volcanism strongly active, hydrothermal vents provided local environments where Miller's considered gases were present, and the energy sources could drive the abiotic synthesis of simple biological molecules near these vents.

Another important source of biological molecules was extraterrestrial synthesis. It is now well established that amino acids, hydrocarbons, and other biological relevant molecules can form in space. Some amino acids of non-terrestrial origin (not synthesized by any terrestrial organism) have been found in meteorites. Since they were not produced on Earth, their manufacture must have taken place elsewhere. Astrophysical research confirms that interstellar clouds contain precursors of various biological molecules (Fry 2000; Maynard Smith and Szathmáry 1999; Maurel 2003; Gribaldo et al. 2007) and cosmic rays or other energy sources can drive their synthesis. Not only that; if this phenomenon was undoubtedly relevant at the dawn of our planet, it must be emphasized that it continues to happen. Even now, as you are reading these lines!

Huge quantities of water (even, a volume corresponding to that currently present on our planet, would have been produced transported to us by comets; Fry 2000, Chap. 7) and biological molecules could have delivered on Earth by meteorites and comet fragments, around 4 billion years ago, contributing to the so-called "primordial soup" of texts that deal with the origin of Life.

By this time, Earth's oceans already contained a rich mixture of biological molecules, including lipid chains, amino acids, bases, sugars, and/or their precursors.

Science now agrees that Earth possessed an impressive pool of biologically essential elements, synthetized in entirely abiotic condition. The first steps towards Life in a barren planet were thus possible. But the most challenging steps were yet to come!

1.3 From Biological Molecules to Living Beings: A Dead End?

We have seen that the synthesis of biological molecules in the laboratory is possible under appropriate conditions, including the presence of chemical elements and energy. However, NO LABORATORY EXPERIENCE has ever allowed the spontaneous evolution of a protoorganism from the constituents of the primitive soup.

It is often said that, given enough time, even extremely improbable events could occur. However, the formation of a functional protoorganism through random collisions in a mixture of precursors contradicts modern scientific principles of causality. Not only that. This failure has even revived support for divine or supernatural explanations.

Scientists, however, developed hypotheses to explain the transition from non-living to living systems.[12] Two main models exist:

1. Metabolic hypothesis: emphasizes formation of a particular chemical environments equipped with catalysts (including mineral or simple organic molecules) that facilitate necessary chemical reactions.
2. Genetic hypothesis: emphasizes the replication of information, favouring the formation of nucleic acids and self-replicating molecules.

In the light of current knowledge, what would be the most correct way to go in order to formulate a satisfactory theory? Things are not simple at all. Proteins (biological enzymes), under DNA control, perform coherent processes within well-defined boundary (e.g., cell membranes, formed by lipid molecules) that isolate and protect the contents while allowing material and energy exchange. The system only works when complete; missing component, such as enzymes, prevent functionality. This creates a circular problem: DNA requires proteins to express genetic information, while proteins require DNA to function.

This problem was partially resolved in 1989 with the discovery of ribozymes (by Thomas Cech and Sidney Altman, Nobel Prize 1989), RNA molecules capable of enzymatic activity (Fry 2000).

[12] Numerous works have been written to tell, explain and present the different theories proposed regarding the transition from the unliving to the living (e.g., Maynard Smith and Szathmáry 1999; Maurel 2003; Gribaldo et al. 2007; Collective Work 2008, 2013a, b; Lane 2015). A volume that I consider essential, because it synthesizes all the theories of the time, is that of Fry (2000). This is without doubt, an interesting reading to have a more complete historical picture than that reported in these pages.

At first, ribozymes proved capable catalyse reactions, such as RNA cleavage and recombination. Subsequently, other discoveries followed, confirming that RNA (at least some particular sequences of this molecule) is able to exert some catalytic activities. One of the most amazing examples is ribosomal RNA (rRNA). Ribosomes are the apparatus where proteins are manufactured: it is there that amino acids are added, one after the other, to form more or less long chains. Ribosomes are made up of two sub-units, one larger and one smaller, both of which include RNA (the rRNA mentioned above) and some protein sequences (relatively short chains of amino acids). Today we know that the RNA of the largest sub-unit is responsible for the catalytic activity for the formation of the peptide bond between amino acids and not the protein component, as one might have imagined (Forterre 2007).[13]

So here is that the problem of the intermediate step seems to have been solved, thanks to the discovery of molecules endowed with both the properties of proteins and nucleic acids.

As time went by, an idea took hold, that of an "RNA World", which would precede the current biological reality. This hypothesis propose that RNA initially performed both roles: storing genetic information and catalysing reactions. Later, evolution favoured DNA for information storage and proteins for catalytic activity. RNA persists in other roles (like rRNA) but no longer in those described above (Fry 2000). It is indubitable, in fact, that DNA, thanks to its particular double helix configuration, is able to store biological information much better and, at the same time, can more easily allow the transcription of RNA and the duplication of the molecule. On the other hand, proteins possess higher plasticity to play the role of natural enzymes. Their ability to fold and protect the active site makes them extremely specific, therefore more effective and versatile than the particular needs of each individual chemical reaction.

However, despite these discoveries and the importance of RNA in the world of Life (they are precisely the tRNAs, particular strands of this molecule each linked to a different amino acid—a particular tRNA corresponds to a single amino acid—which, properly activated, work as a precursor of proteins; moreover, it is enough to remember the ribosomes, which we have just talked about and the importance of rRNA for the production of proteins), challengers remain, even with the introduction of ribozymes: RNA, is unstable

[13] P. 109 and text framed on p. 151 of this paper.

and unable to form stable helices for storing information, and if a "RNA World" existed, why we do not find any traces in modern biology[14]?

Moreover, even if we had solved the circular problem (does the program or the chemical system come first to express it?), this solution remains overly sophisticated: no primitive soup evolves spontaneously into an "RNA World"! How, then, could such a stage have been reached? We have not yet managed to overcome this impasse.

In previous debates, I overlooked an important detail concerning chemical reactions. Life depends on the interaction between a series of biological molecules that form the "program" (the software regulating the living organism) and those that facilitate and accelerate chemical reactions, including those related to metabolism and the production of the very molecules mentioned above. This statement, while correct, is incomplete: something is missing. But what exactly?

Let us return to our intuitive definition of the living being; it must "reproduce" and "eat." The latter ability is undoubtedly essential, not only to provide the material needed for growth and reproduction, but also to supply the energy required to carry out biological functions. In fact, chemical reactions fall into two categories: those that occur spontaneously (also called "exergonic reactions") and those that require energy to proceed (the "endergonic reactions"). Catalysts (enzymes) can facilitate the first type, but although necessary, they are not sufficient for endergonic reactions, which require an external energy source.

Most metabolic transformations are endergonic and therefore require energy. All current organisms, unicellular or multicellular, possess internal biological "batteries"[15] capable of supplying the energy necessary to power their reactions. The production of these batteries is ensured through feeding,

[14] Actually, the phenomenon known as RNA interference may be a faint remnant of the "RNA world", though it is still too early to conclude. This is a complex phenomenon; interested readers can consult Jablonka and Lamb (2006, Chap. 4, pp. 332–333).

[15] The biological molecule par excellence capable of providing energy to endergonic reactions is ATP or adenosine-5′-triphosphate. By losing one or more phosphate groups through hydrolysis, this molecule produces a more or less energetic exergonic reaction. ATP is also one of the precursors of RNA; it is the molecule that possesses the energy to bind the nucleoside adenine to the rest of the chain (the other precursors are the 5-triphosphates of uracil/thymine [the latter for DNA], guanine and cytosine: they too can serve as "batteries" to store biological energy). The key to energy reserve is linked to phosphate groups. Finally, I recall the nicotinamide adenine dinucleotide phosphate, abbreviated in NADP, another molecule capable of providing energy to endergonic reactions. Its reduced form, charged with energy, is called NADPH or NADPH2 or even NADPH + H^+.

which allows them to recharge.[16] This is always been true for living beings, as we understand and define them. But what happened before?

We have already recognized the importance of energy intake in the first stage of the transition from non-living to living matter: the production of biological molecules. Whether this process occurred on Earth or in space (or both), an appropriate energy source was always required. The need of energy has always been considered in theoretical models and experimental designs.

But what type of energy could have driven the elements of the primitive soup through subsequent stages? Most scientists investigating the origin of Life, at least in its earliest steps, have assumed that the energy source was intrinsic to the components of the soup itself. Such a mixture of biological molecules contains precursors of major biological constituents as well as molecules needed for nutrition, such as sugars. However, this energy source would have been created under specific conditions, probably preceding the stages we are examining. Later, production either ceased or was greatly reduced, because these energy sources could have interfered with subsequent stages. Consider, for example, the consequences of a meteor shower impacting precisely where the chemical processes required for the synthesis of increasingly complex biological elements were occurring. Or imagine the heat from a hydrothermal spring, which, while promoting the formation of simple biological compounds, simultaneously inhibits more complex synthesis processes.

Thus, the processes of interest likely occurred once the phenomena that produced the primitive soup in the ocean have subsided.[17]

Once formed, this soup was not renewable (or only minimally so). Depletion of this reserve would have ended the processes leading to the emergence of Life. Moreover, to properly utilize these molecules as food, the system would have required appropriate enzymes to process them. Again, there is no evidence to suggest that such enzymes could have appeared spontaneously. The energy source necessary for the evolution of organisms from the primitive soup cannot be found within it, except under extremely particular conditions.

Have we perhaps reached dead end?

---

[16] The degradation (oxidation) of a glucose molecule produces the equivalent of about thirty ATP molecules (in the form of ATP and NADPH). Various ATP and NADPH molecules can be produced during the light phase of chlorophyll photosynthesis. However, I would like to remind you that food is not just a source of energy for the body. It is also necessary to enable the organism to obtain appropriate chemical elements (nitrogen, carbon, particular amino acids, etc.).

[17] Another way of looking at things is to separate the production of biological molecules geographically and the next steps. However, this does not change the following speech.

1.4 Mario Ageno's Proposal

Researchers who dealt with the transition from the non-living to the living were well aware of the problems involved and, of course, proposed various solutions. These can be found in the vast literature on the subject, both scientific and popular.[18] However, here I would like to discuss the hypothesis proposed by Mario Ageno, the same researcher whose definition of a living being we encountered in paragraph 1.1. I wish to present Ageno's ideas because, although they were published in Italy during the 1990s, they have never reached a broader audience.

Despite the time has passed that has passed and the discoveries made in recent years, Mario Ageno's proposal, in my opinion, still offer valuable insights for those seeking a scientific explanation for the emergence of Life.

I will not address particular issues, as the asymmetry of the biosphere or the availability of certain chemical elements. These topics, while interesting, are too specialized for a general popular work focused on the broad outlines of the remarkable story of Life. In any case, Ageno (1991) discusses them in his book, which I recommend consulting for further information and its rich bibliography.

Let us now move on to the most interesting aspects of this proposal. In its initial part, it does not differ greatly from classical hypotheses. It involves the formation of the primordial soup containing specific molecules and biological precursors are found, just as in most developed theories.

Mario Ageno situates the evolution of the soup in a particular environment: a lagoon, a sheltered habitat, though still in contact with the open sea, under a relatively thick layer of water, several meters deep, so that ultraviolet radiation from space could not destroy the biological molecules formed. At that time, the modern atmosphere capable of shielding us from high-energy radiation had not yet formed. However, the depth should not be excessive: visible sunlight still needed to reach the bottom, where the chemical elements of the soup were present. A lagoon would have allowed the substances to transform without being disturbed by external phenomena.

As mentioned, light will play an important role in Ageno's hypothesis, which we will address later. For now, let us follow the sequence of events. In this particular habitat, under the proposed conditions, the layer of hydrocarbons in the primordial soup disaggregated into lipid pockets (or spherules). What do these new entities represent?

[18] See footnote 12 of this chapter. In any case, the most complete book on the subject is, in my opinion, that of Fry (2000). However, although remarkable, the author, not knowing Italian, has not had a chance to learn the proposal described in this paragraph. Also see Mieli et al. (2025).

Think of an oil slick on water. The oil does not mix with the water. Movements of the liquid may break up the oil slick, but its molecules will not disperse; instead, they form small spheres in the water, similar to soap bubbles. This is how the lipid sacs form.

It should be noted that many lipid molecules (which are made up of fatty acids) have an end that does not like water at all (which is why it is called "hydrophobic") and another that, on the contrary, tolerates contact with this liquid (this second end is called "hydrophilic"). If the concentration is not excessive, the molecules are soluble in water. But if this increases too much and exceeds a certain critical value, the lipid molecules form a double-thick layer on the liquid. The hydrophobic ends will have a tendency to arrange themselves opposite each other, while the hydrophilic ends are all arranged in contact with water (Fig. 1.5). This explains the formation of an oil stain in an aqueous liquid.

Lipid hydrocarbons were an integral part of the primitive broth. Their concentration allowed the formation of large oil layers in the liquid, which currents could be broken into lipid pockets (Fig. 1.5). These pockets could reach the protected environment of the lagoons. Let us pause to examine

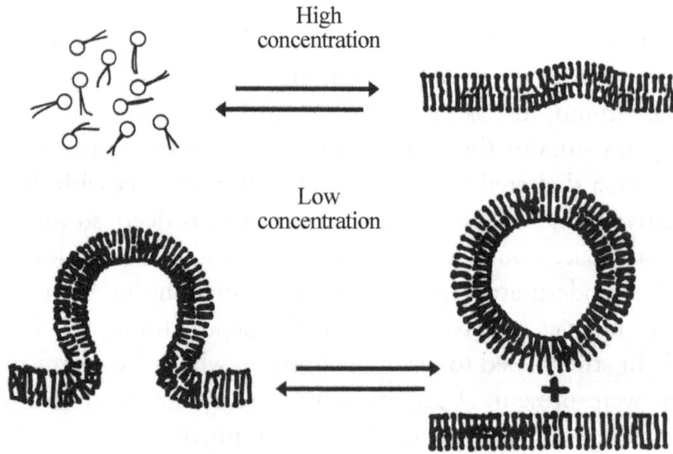

Fig. 1.5 If the critical concentration is not reached, the lipids are soluble in water (top left). But when the critical contraction is exceeded, a lipid bilayer is formed, with the hydrophilic (water-tolerant) ends of the lipids facing outwards (these are the rounded heads of the molecules) while the hydrophobic part (which does not tolerate it) is contained inside (these are the elongated tails of the molecules). In the lower part of the figure, the formation of a lipid sac is shown, due to the movements of the liquid. Redrawn from Ageno (1991)

the properties of these spherules, as they are crucial to understanding what follows.

Lipid sacs have very specific properties; they isolate their internal compartment from the external environment. Only fat-soluble substances (able to dissolve in fatty acids) can penetrate the sacs (for example, with CO_2). In contrast, all polar substances (water, ions, and most molecules of the primordial soup) cannot enter. Therefore, elements inside the sacs remain at the time of formation remain trapped, while non-fat-soluble molecules cannot enter. The internal and external contents of the spherules cannot be mixed.

But the properties of lipid spherules are not limited to this. Due to the liquid movements, two or more sacs may collide and fuse into a larger sphere, a process called "pinocytosis". The contents of the original sacs remain intact inside the new structure because there is no mixing between internal and external environment. A sac can also split into two smaller ones, dividing its contents between them, again without mixing with the outside.

This means that during pinocytosis, no new material enters the sacs, nor can it escape. Thanks to waves and currents movement, lipid pockets, with their precious contents, together with the primordial soup components outside, are pushed into the lagoon, a few meters below the surface, not far from the bottom.

Other authors, particularly proponents of the metabolic hypothesis, had already proposed similar scenarios (without lagoons), where protoorganisms could evolve from individual pockets containing a complex chemical apparatus. However, this process has not been successfully reproduced in laboratory experiments.

This is where Ageno's proposition diverge from classical hypotheses. Let us examine how.

First, it is incorrect to consider the evolution of the internal contents of a single lipid bag. Because the internal and external materials cannot mix, the system should be seen as two environments: the external one, and the collective internal environment of ALL the lagoon's pockets. In fact, since the internal material can never mix with the outside (and vice versa) due to the properties of pinocytosis, what happens is that the molecules contained in one sac (say, sac "A") may eventually contact inside another sac (sac "B") through fusion. This means that a reaction initiated in one pocket ("C") can continue in another ("D") when their contents merge. All this, in a habitat that always remains separate from the external environment of the lagoon. This way, the system can continue to evolve more easily.

The fact of considering the whole of the internal cavities of all the pockets of the lagoon as a single environment, as opposed to the external one, avoids us having to consider a single pocket fate that will practically never spontaneously evolve towards a coherent configuration. Instead, considering the set of ALL pockets' internal environments, the probability that a series of coherent chemical reactions can be established becomes much more important.[19] Let us remember that in our system, one sac can fuse by pinocytosis with another and share its contents with it. The molecules present in a particular sac are, therefore, susceptible, sooner or later, to meet those inside any of the other spherules of the lagoon.

However, this alone does not solve the problem of energy supply. Molecules of the primordial soup would have been an insufficient energy source, and their metabolism would have required suitable catalysts. This challenge remains if we focus solely on the internal contents of individual sacs. How does Ageno address this?

Before continuing, let us understand the dietary strategies of living organisms, current or fossil. They fall into two broad categories: heterotrophs, which obtain food from their environment, and autotrophs, which synthetize their own food from simple precursors (e.g., water and CO_2). Animals and many microorganisms (as well as amoeba or bacteria) are heterotrophs, while plants and cyanobacteria are autotrophs, able to use light to synthetize glucose.

Be careful, plants do not simply feed on water, CO_2 and light. Although the process of these elements also makes it possible to recharge, in part, the body's chemical batteries (those that are then used in all endergonic biological reactions[20]), the final product of all photosynthesis reactions is glucose. Once produced, it is used, through the same metabolic reactions shared by heterotrophs, for purposes identical to those of other organisms. Plants, therefore, are perfectly capable of making their own foods that they will later digest like animals and microorganisms. In addition, this process allows them to produce additional energy.

Certain microorganisms (particularly those in current hydrothermal vents) are chemoautotrophs, generating glucose from chemical reactions.

Let us now consider the energy sources available at the time of the transition from non-living to living. At the time, the only omnipresent, inexhaustible energy source was the Sun, which even today, is responsible for Life on Earth. The day this star goes out, all plant Life we know will disappear,

[19] The whole process must no longer take place in an isolated bag. It is sufficient to start in a particular sphere; the rest will be done when the contents of the first bag are brought into contact (by pinocytosis) with those of another one in which the elements are present for continuing the process (Ageno 1991, paragraph 7.1).

[20] See footnote 15, in this chapter.

pushing all animals that feed on it and all beings that depend on plants to extinction. The only organisms that will be able to survive the catastrophe are those that thrive using energy sources independent of solar energy. These sources are fed by processes typical of the inner layers of our planet, which produce, among other things, underwater hydrothermal vents (let us not forget about this particular environment, because we will have to talk about it later!).

For Ageno, solar light powered the processes leading to Life. Let us now see, quickly, how this was possible.

Plants and cyanobacteria have mechanisms to capture light energy, converting it into chemical energy stored in particular molecules,[21] or used in particular metabolic pathways to transform CO_2 into glucose. Plants and cyanobacteria are capable of photosynthesis.

In the following pages, we will limit ourselves to the process of converting light energy into biological energy. The production of glucose goes beyond the scope of my book and does not Ageno's proposal on the transition from the non-living to the living.

The photosynthetic mechanisms that interest us are sophisticated and also particular. They are located within the thickness of a lipid membrane, of the same type as the double-thickness spherules seen in the previous pages. The peculiarity of the green plant system is that it includes two different light-absorbing centres. They work in unison and are called "Photosystem I and II" (Fig. 1.6).

Each photosystem contains chlorophyll a. If it receives an appropriate photon (a particle of light),[22] it loses one electron, which transferred to another molecule.[23]

To understand how the light-sensitive photosynthetic mechanism works, let us examine what happens, starting with photosystem II (which was so

---

[21] These molecules are involved in all endergonic biological reactions. They have already been mentioned in footnote 15 of this paragraph. The most important of these molecules is, without doubt, ATP, which can provide energy through the loss of one or more phosphate groups. Regarding the production of glucose, this happens during that phase of photosynthesis that does not depend on light (and therefore can take place even in its absence), thanks to the energy generated in the previous phase, called "luminous phase". And it is precisely the luminous phase (and only this) that we will be interested in later.

[22] Each photon has an amount of light energy that depends on its wavelength. Chlorophyll is sensitive only to certain wavelengths, corresponding to blue and red light, with variations depending on the particular type of chlorophyll being considered.

[23] Of course, the lost electron must be regained again, otherwise the chlorophyll would remain unbalanced. In green plants, the electron is released by tyrosine Z (tyr Z), whose task is, rightly, to supply the electrons lost to chlorophyll. To become stable again, the tyrosine must pick up another electron and take it from the water (H_2O). The process which enables this action (passage of electrons to tyrosine) releases gaseous oxygen (O_2). And this is precisely why the photosynthesis of green plants (and also that of cyanobacteria) is able to produce this element.

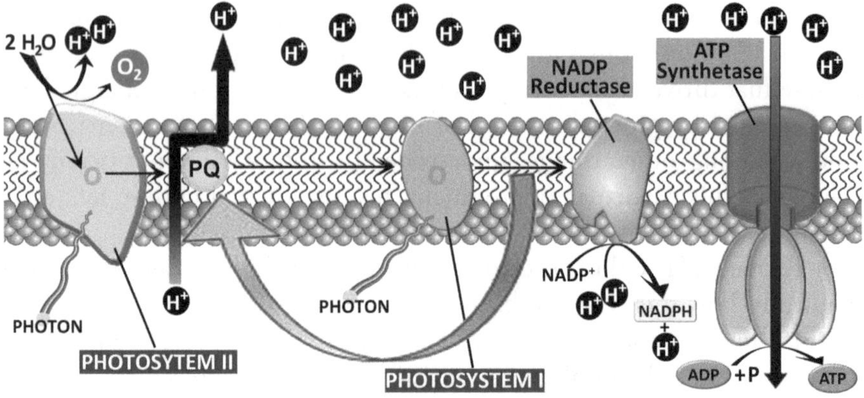

Fig. 1.6 Diagram of the photosynthetic apparatus of green plants (only the structures shown in the text appear). The components are housed within a bilayer lipid membrane, the thylakoid membrane. The two light-absorbing centres are indicated as "Photosystem I and II", the plastoquinone as "PQ", the path of the H^+ ions, which cross the membrane, is visualized by the large black arrow, on the left. Most of the molecules that make up the transport chain are not included in the diagram. The decomposition of water into H^+ ions and oxygen (O_2) is shown in the upper left. The formation of ATP, the main biological molecule capable of storing energy, from ADP and phosphate groups, can be seen in the lower right, under the structure of ATP synthetase. The large light grey curved arrow visualizes the path of cyclic phosphorylation, described in the text. Redrawn from Purves et al. (2003)

called simply because it was discovered after the other, baptized photosystem I).

An appropriate photon excites the chlorophyll a of system II, which loses an electron. To regain its stability, this molecule must find the lost electron(s). It obtains them thanks to the decomposition of water (Fig. 1.6) which, in turn, produces H^+ ions, gaseous oxygen (O_2),[24] plus electrons that will be used to restore the balance of chlorophyll.

But what happens to the electrons released by chlorophyll? They move to another element which, in turn, quickly transfer them to another molecule. In this way, the electron moves from one molecule to another, along a chain characteristic of any type of photosynthesis. And it is thanks to this transport that part of the plants' biological energy is produced. A simplified diagram is shown in Fig. 1.6. What is important is to understand the global mechanism: due to light, an electron is torn from chlorophyll and moves along a particular

[24] This is the reaction that produces the oxygen released by green plants (and cyanobacteria). This is due to the particular mechanism of photosynthesis with two photosystems, with water supplying electrons. For details, see also the previous footnote.

"path". The light energy of the photons is then transformed in this process and can recharge the biological batteries.

Along the transport chain of electrons, green plants and cyanobacteria, the electron lost by photosystem II passes through a particular molecule, plastoquinone (PQ). The latter is able to literally "swim" in the thickness of the membrane and can reach the two boundaries. But the main property of plastoquinone is another: when it receives electrons, it must also acquire an equal number of H^+ ions. These are taken in the aqueous environment beyond the membrane (Fig. 1.6, bottom). Later, when the PQ transfer the electrons to the next molecule, it also gets rid of the H^+ ions. But this time (and this is the crucial point), the electrons are released on the opposite side from where they were taken (Fig. 1.6, top)! In summary, one of the two regions separated by the lipid membrane of photosynthesis (that of green plants, called the thylakoid membrane, is found within particular organelles[25] the chloroplasts), thanks to the system just described, is depleted in H^+ ions, while the other is enriched by them. Ions, on their own, cannot cross the membrane: to do so they need the PQ (see Section "Quinones and Their Properties", in Appendix).

Before we continue to reflect on the importance of this fact, let us follow the path of our electrons to the end of the photosynthetic chain. When they leave the PQ, the electrons continue their journey until they reach photosystem I which, like the previous one, has lost one or more electrons due to as many photons. Thanks to the sequence described, photosystem I regains equilibrium, after having lost electrons due to light. But where do the electrons he lost go? Continuing a similar path to the previous one, from one molecule to another, they arrive at the "NADP reductase" complex, which is capable of producing NADPH. The latter is another molecule capable of storing biological energy.[26]

But we have not yet reached the end of the surprises..., if you look carefully at Fig. 1.6, you will notice that a certain amount of H^+ ions has accumulated on one side of the membrane (top, in the figure). This accumulation is due to the action of the PQ (in part, it is also due to the fact that water decomposes, producing H^+ ions, to provide electrons to the photosystem II). This electron excess is used to make ATP synthetase, a complex that crosses the thickness of the membrane, work. By allowing ions to pass through it, so as to restore ionic balance on both sides of the membrane,

---

[25] An organelle is a specialized structure located inside a cell. It is bounded by a double-walled lipid membrane, such as those we have discussed in this chapter. Mitochondria and chloroplasts are typical organelles of eukaryotic cells.

[26] See footnote 15, in this chapter.

it acts by producing a particular molecule, ATP, which is the main energy battery of the cell (Fig. 1.7, top).

And that's not all. In the path that starts from photosystem I, there is a variant of the normal path that consists in making the electron travel the same path several times, this is the "cyclic phosphorylation" (shown in Fig. 1.6, by the large grey curved arrow). Basically, the electrons are sent back to the PQ which, subsequently, puts them back on the path to system I. The electron leaves this path and returns through the PQ.[27] The advantage of this variant is that it contributes to the enrichment of H^+ ions, the region that will allow ATP synthetase to function. The latter can therefore produce even more ATP molecules while the number of photons received by the photosynthetic apparatus remains unchanged.

All this machinery may seem complicated and excessive, but the two centres, photosystems I and II, are both necessary because they do not work with photons that have the same energy: those that excite the chlorophyll a of system I are not effective for the chlorophyll of the other.

Now imagine simplifying our system by reducing it to:

1. The cyclic phosphorylation device (suppose we simplify it further, reducing it to some element, including PQ or an equivalent molecule, because there are simpler substances that retain the same properties).
2. Chlorophyll (or another appropriate pigment, i.e., another molecule capable of becoming excited by light).

The minimal system we are interested in is one capable, thanks to photons, of transferring H^+ ions to the other side of the membrane.

Let us now return to our initial lipid pockets and suppose that the molecules we have spoken of before (pigment, PQ or equivalent substance, etc.) could have been imprisoned in the thickness of the membrane, which is statistically plausible under the conditions and at the time considered. The light excites the pigment, which loses some electrons that reach the molecule with the properties of the PQ. This, receiving the electrons (say two), will have to acquire as many H^+ ions (therefore, two ions) in the EXTERNAL aquatic environment. When the molecule returns to its normal state, it will release ions and electrons. This time, however, the H^+ ions will be transferred to the INTERNAL environment of the sac. After a certain time, therefore,

[27] In all current photosynthetic apparatus, the cyclic phosphorylation device seems to be always present (Ageno 1991, p. 275). This leads us to believe that this system is the central nucleus from which the definitive mechanism was developed. This finding not only helps us to understand how this type of apparatus could have established itself, but also comforts us on the validity of the hypothesis issued.

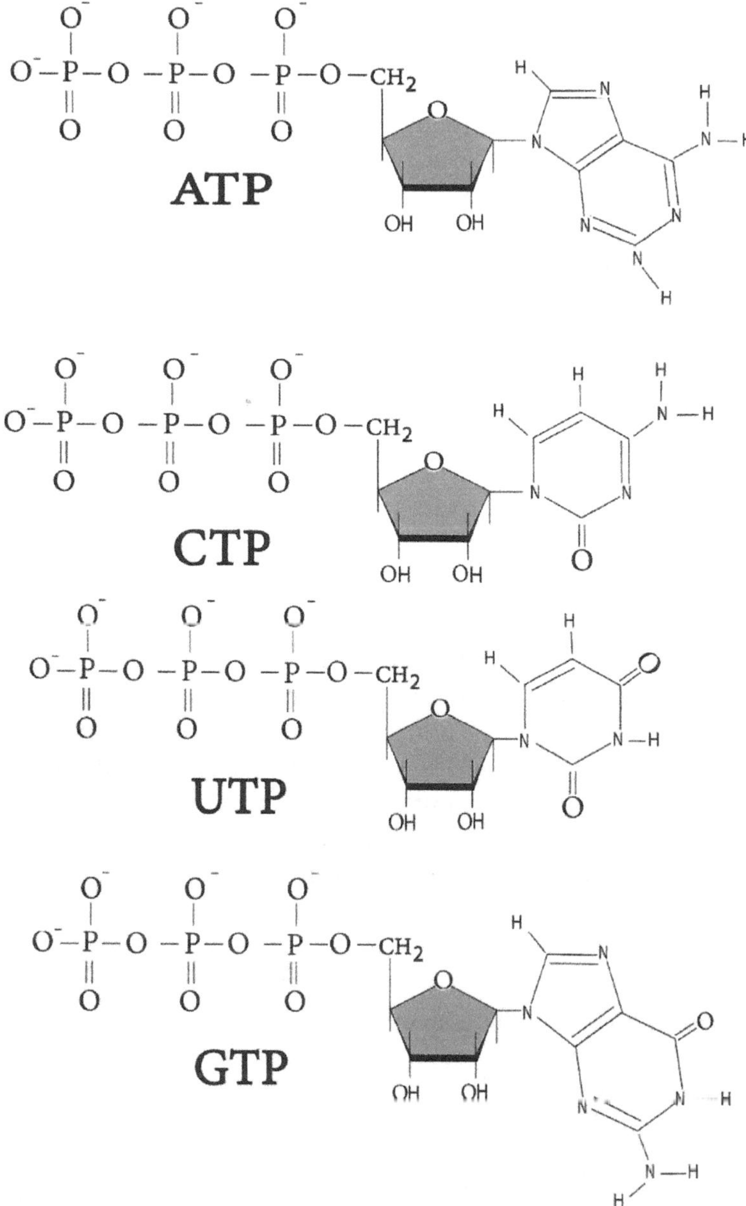

Fig. 1.7 Adenosine triphosphate (ATP), top of the figure. This molecule, the main supplier of biological energy in the cell, is none other than one of the four precursors of RNA. The other three are cysteine (CTP), uracil (UTP) and guanine triphosphates (GTP; to obtain them, simply replace adenine with another base). These molecules, just like adenine triphosphate, can be used as biological batteries. If you want to obtain DNA precursors, simply start with these four molecules and replace the –OH radical, which is bound to carbon 2′ (bottom right of the sugar ring) with a –H and, of course, change the uracil base with thymine. Redrawn from Hebsgaard et al. (2005)

the internal environment will have been enriched with H^+ ions: it will have become acidic compared to the outside.

It is true that the acidity inside the sac also depends on the collisions that it makes with the other spherules and the divisions that follow. Some can lower the concentration of H^+ ions in the bag containing the device, because they can be segregated into other spherules. In addition, it can receive substances that are likely to attenuate acidity.

However, in sufficiently acidic pockets (especially in the one—those—containing the simplified cyclic phosphorylation device) some mono- or polyphosphates (molecules that contain only one or more phosphate groups, the chemical group $PO_3{}^{2-}$, which in acidic environments can exist in the reactive state) can be formed. These phosphorylated molecules, once returned to less acidic environments, can function as biological batteries and release their energy of phosphoric bonds for the synthesis of new substances. In fact, the phosphate groups are responsible for storing the chemical energy used by living beings. The latter, in turn, can be used for endergonic reactions. ATP is the main molecule with these properties. It can provide energy by losing one or more phosphate groups. It can lose up to three.

Now imagine that nucleotides and nucleosides are part of our system.[28] Starting from these precursors, in the presence of protonated phosphate groups, active precursors can be produced (nucleotide triphosphates; Fig. 1.7) which, in turn, allow the formation of DNA or RNA strands. These same precursors coincide with the most important molecules for the storage of biological energy (and all this, most likely, is not at all accidental!).

In the presence of the appropriate material and with the appropriate forms of energy (the elements of the primitive soup contained in the lipid pockets and the batteries made up of the molecules provided with phosphate groups) the production of a nucleic acid filament becomes achievable, without resorting to any living being (Ageno 1991, Chap. 7).

The proposition put forward by Mario Ageno would therefore allow, starting from the scenario of the primordial soup, to obtain two important results: (1) production of energy molecules to drive chemical processes; (2) abiotic formation of nucleic acid strands.

Is this scenario plausible? Is it possible to form the mechanism transferring electrons through the membrane of the sacs, enriching the internal environment with H^+ ions? The presence of molecules with double bonds between two carbon atoms in the soup does not pose any difficulty. The equivalent of

[28] On the possibility of synthesizing RNA nucleotides see Powner et al. (2009) and Ritson and Sutherland (2012).

PQ may therefore have been in the right place at the right time (thus being trapped in the lipid double-walled membrane).

That the synthesis of chlorophyll precursors was entirely possible. On the other hand, this molecule would be very old: some photosynthesizing organisms, capable of using water as electron donors (organisms that, therefore, must necessarily possess chlorophyll) have been found in rocks about 3.5 billion years old Warrawoona Group (Australia; Schopf and Parcker 1987). This implies that such a molecule, or an appropriate precursor, was certainly present. Regarding amino acids and the bases of DNA and RNA, some of them and/or their precursors were part of the set of biological substances present in the soup.

From this it can be deduced that Ageno's hypothesis, at least on a purely theoretical level, is perfectly valid. Not only that, but it allows us to consider as possible the production of polynucleotides, strands of DNA, RNA or even a mixture of the two types of molecules, all obtained in an environment perfectly devoid of living beings. This would allow us to reach the threshold of the "RNA World", so dear to many modern researchers.

Of course, Ageno's idea does not stop at this stage. Not believing too much "RNA World", he assumed an early separation, from their first appearance, between DNA and RNA. In fact, according to Ageno, the first is the only one that can guarantee the formation of a very stable molecule to function simultaneously as the seat of the genetic heritage and as a program.

His scenario even contemplates the formation of polypeptide chains, obtained from hybrid molecules, including an amino acid and an RNA sequence, the future tRNA, which are the precursors from which proteins can be synthesized. Since, thanks to the process described, nucleotide chains could have existed, it is possible that, under these conditions, proto-tRNAs could have interacted with strands of these molecules (especially RNA, produced using DNA as a template) by putting themselves in the right position to allow two amino acids to bind correctly.[29] Subsequently, through the same mechanism, a third amino acid could have been added to the chain, then a fourth, and so on, until an oligopeptide sequence (including a few amino acids) was obtained. It would have been the first example of an amino acid sequence formed under the control of a molecule consisting of nucleotides (Fig. 1.8).

The unique correspondence between a particular RNA molecule with a single amino acid would have allowed the birth of the genetic code. How could it have been realized? Starting from trial and error, based on products

[29] In order to deepen these processes and also all the phases of the hypothesis proposed by Ageno, I refer you to his book (Ageno 1991, Chap. 7). For an evolution of his theory, see Mieli et al. (2025).

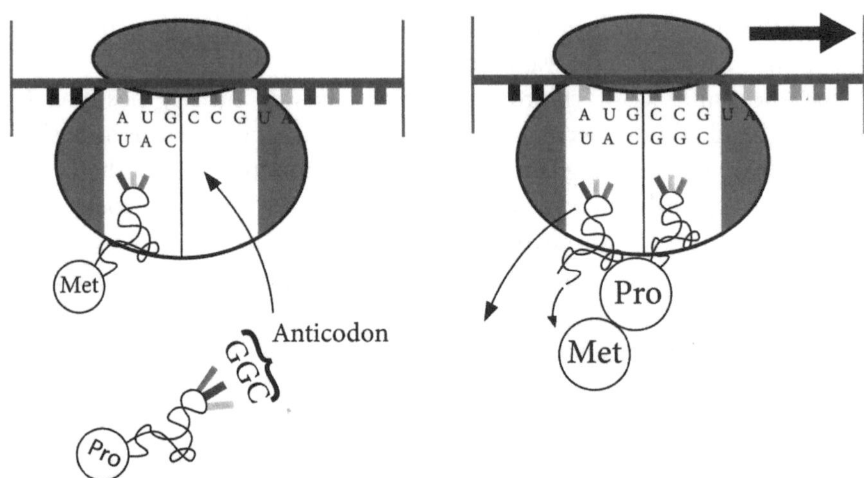

Fig. 1.8 Formation of a peptide sequence from the sequence of messenger RNA (mRNA) codons on the ribosome (in grey). On the left, a tRNA with the amino acid methionine (= Met) has associated with the corresponding codon. A second tRNA is coming. Its anticodon (the complementary sequence to the codon; in this case, it is "G-G-C", i.e. guanine-guanine-cytosine) corresponds to the amino acid proline (= Pro). On the right, the second tRNA associated with the mRNA and the two amino acids were able to bind to each other. The first tRNA can move away from the ribosome while the latter is ready to move along the mRNA (to the right, in the direction of the arrow). In this way, the second tRNA, with the two amino acids bound together, is in the position of the first and, to its right, there will be space to accommodate a third tRNA. In this way, the cycle repeats. Redrawn from Purves et al. (2003)

obtained from the simple random mechanism. In fact, the regulated synthesis of oligopeptide sequences would have allowed the production of very simple enzymes. These could have favoured particular chemical reactions instead of others. In this way, a first "evolution" could have been produced, capable of acting on the starting products and molecules. The selection would have favoured the synthesis of peptide sequences that facilitated the production of those DNA molecules that served as a model for the manufacture of the above sequences and here, therefore, our chemical system begins to be regulated by a program draft! The consecration of a genetic code (link between an tRNA and a particular amino acid) completes our scenario.

As soon as all these processes take place within the same sac (which has now become capable of functioning without the need to interact with the others) we can begin to talk about a protoorganism. I remind you, as I have already explained in paragraph 1.2, that it is the protein/amino acid interaction that allows Life to be realized according to the definition given.

1.5 Where Does Life Come from?

Ageno's hypothesis, published in 1991, remains yet unproved experimentally. However, they are very promising and can shed new light on various stages of the transition from non-living to living. They may even allow us to lay the foundations of a comprehensive theory on the subject. Time must be left for future research that will be able to decree the validity of the suggestions or correct them.

Over time, although Ageno's ideas never crossed national borders, there was some change in the world of research on the origin of Life. Regardless of his formulation, various researchers suggest that the first living things could have been autotrophs, even if the hypothesized mechanisms had nothing to do with cyclic phosphorylation.

I have already mentioned (paragraph 1.2) hydrothermal vents as one of the sources of production of biological molecules. Let us now resume the subject. The ecosystems that thrive in these environments are completely independent of the Sun's energies (remember, however, that the lifespan of an isolated source is ephemeral compared to that of a star!). Heat and hydrothermal processes are able to feed, on their own, a complex food chain that includes simple microbes up to colossal worms, crustaceans and fish.[30] Some scientists have put forward the hypothesis that these places, sheltered from any possible cosmic attack, could have hosted the cradle of the first forms of Life on our planet. In fact, it has been said that the first molecules could have formed in the iron sulphide-coated alveolar wall of hydrothermal vents. These, later, would have aggregated giving rise to increasingly complex compounds, until they formed macromolecules: for example, proteins or nucleic acid strands (such as RNA).

And the engine to carry out all these processes? The heat emanating from the lower layers of the Earth, capable of rising to the surface. Evolution would have progressed to the formation of a cellular system wrapped in a lipid membrane. Therefore, until the emergence of a real living organism.[31] Traces of microorganisms susceptible to having lived based on metabolic pathways such as those that in microbes that thrive in underwater hydrothermal vents

[30] The reader interested in the fauna of deep hydrothermal springs can consult book by Desbruyères et al. (2006). However, this work only concerns animals living and/or frequenting these environments: micro-organisms are not treated.

[31] See the article by Paetsch and Wehrmann in Collective Works (2013a), and the one by Sojo et al. (2016).

would actually have been found in sediments as old as the first traces of Life.[32] According to David Wacey and his collaborators (Wacey et al. 2011), in Australia, these organisms would be even older than the stromatolites found at the same site, but in different layers.

Is it possible, then, that the first organisms were chemoautotrophs, having developed near hydrothermal vents? This is a perfectly plausible hypothesis, but we should imagine appropriate mechanisms to allow us to validate all the stages conceived, because the discussion just made on cyclic phosphorylation does not seem to fit this case as well.[33] It can be used in the case of photosynthesis (of all types of photosynthesis, not only from that carried out by green plants and cyanobacteria), but it is difficult to transpose in case of chemosynthesis. On the other hand, fossil traces, even very ancient, of photosynthesizing organisms capable of supporting the validity of Ageno's hypothesis, have actually been found and, in my opinion, his remains the best scenario to explain the emergence of living beings.[34]

But is it possible that Life appeared more on Earth, under different conditions? We will return to the subject in the next chapter, after talking about the oldest known living organisms.

But before concluding the chapter dedicated to the transition from the non-living to the living, I still have to discuss one last hypothesis, which is sometimes raised when talking about the beginnings of Life on Earth: the possibility that it came from another planet, via an asteroid or a comet.

I have already mentioned (paragraph 1.3) that no one, in the past, has ever been able to explain, in a convincing way, how Life was formed on our planet. This failure has led some scientists to think that the approach was not the correct one: rather than being able to elucidate the formation of Life on Earth, it was easier to move the problem elsewhere and believe in an appearance on another planet. Let us not forget that biological molecules can indeed form in space, before reaching a planet. Why, then, could not the first living beings have taken the same path?

---

[32] In this regard, read the articles by Rasmussen and Shen (Rasmussen 2000; Shen et al. 2001). The latter refers to traces dating back more than 3 billion years in Australia, which testify to the existence of micro-organisms capable of reducing sulphates since this time.

[33] For other criticisms of the hypothesis of life appeared in hydrothermal springs see Forterre (2007, Chap. 4, Annex 2). I want to mention also two excellent books by Lane (2010, 2015) who defends this hypothesis, using elements which may prove decisive in future research.

[34] About such tracks see, for example, Taylor et al. (2009, Chap. 2) and Tice and Lowe (2004). According to some researchers, the traces of photosynthetic activity could be even older than mentioned in the last article cited above (Rosing and Frei 2004).

Having eliminated the possibility that Life could have formed in the cosmic vacuum, the hypothesis that it was generated on another planet remains. From there he could have moved to Earth. But from where exactly?

The planet that has found the greatest success from this point of view has been Mars. Even today, many space projects try to find out if this planet does not hide some form of Life (or if this has not been present in the past). Its proximity to Earth makes Mars an excellent candidate for the origin of living beings and for the dispersion towards our planet (do not forget the fascination that Mars has always exerted, for example in science fiction: the term "Martian" now reserved for any alien being, was originally created in honour of the alleged inhabitants of Mars).

But is the hypothesis of a Martian origin of Life really credible? So far, all tests to prove Life on Mars have given a negative result. However, this does not prevent us from hypothesizing that, in the past, at a time more or less close to that in which Life made its appearance on Earth, Mars was not really inhabited by living organisms. The most compelling arguments are found in the book by Meinesz (2008), professor of biology at the University of Nice-Sophia Antiopolis.

The author is a partisan of the thesis stating Life appeared on the "Red Planet" and arrived on ours traveling on a meteorite, a fragment of Martian soil detached from the planet and hurled into space due to the impact with a celestial body. From space, the fragment would have arrived on Earth and would have allowed the seeding of the first forms of Life on our planet. Is this realistic?

Indeed, we know of various meteorites of Martian origin. The chemical composition makes it possible to identify their origin and also the age of formation. Among all these, one deserves a special mention: the ALH84001 meteorite formed, approximately, 4.5 billion years ago (which makes it one of the oldest known rocks of planetary origin, Earth or other planet). Specialists think that this rock was torn from its place of origin 16 million years ago (My). After wandering through space for a long time, it made landfall on Earth around 13,000 years ago (Fry 2000, Chap. 14).

As soon as origin and age were defined, we moved on to examine its content. Some traces inside the meteorite suggest "stigmata" of living organisms. What is it about? Some microscopic and elongated corpuscles have come to light. Apart from the shape, what intrigued the researchers who dealt with the meteorite was the presence of carbonate globules. These, in fact, contain small magnetized crystals: real miniature magnets.

Corpuscles of this type, are well known in living beings, their task is to help the orientation of organisms. Inside, whether they are simple bacteria or more complex organisms, these magnets are arranged in single file and each element is separated from the previous and the next by a very fine membrane.[35]

This discovery, in the bowels of the ALH84001 meteorite (the fact of finding them well nested inside the rock made it possible to exclude terrestrial contamination) posed the problem of a possible extraterrestrial biological origin to explain these structures. Some scientists were quick to speak out in favour of the ancient Martian Life form. Especially since another meteorite, called Nakhla, has begun to intrigue researchers, again about the possibility of ancient forms of Life on the red planet.

Currently, all these traces, which would suggest ancient Life forms on Mars, remain hypothetical, although it is not easy to consider them as trivial artifacts due to mineralization. The problem, therefore, is still open.

However, even if the biological origin of these finds were to be demonstrated, it still remains to be explained how they managed to reach our planet while still alive.

Note that terrestrial spores and microbes have been found in space; even on Mars! They were transported from our planet, due to insufficient sterilization of the devices we sent into space to study other celestial bodies (Collective Work 2006). To the current state of our knowledge, none of these organisms is capable of reproducing. This means that we have already begun to pollute the cosmos with our terrestrial microbes!

But let us get back to our problem. In order to migrate to another planet, different from the one of origin, a living being must be able to withstand a journey in space that can be very long. The meteorite ALH84001 has been in space for almost 16 million years (My); an exploit that no astronaut is capable of!

However, we should not underestimate the capabilities of living beings; even those of simple microorganisms. A microbe or a spore can find protection even inside a trivial pebble. Inside, it will be sufficiently protected from all the dangers that a journey into the space vacuum can present. And if the microbe (or spore) in question is able to endure, in a "quiescent" state, a long enough journey, it could also be able to colonize another planet.

How long can a microbe or spore go into a quiescent state? Although the scientific literature reports cases of several million years, scepticism in this regard is still strong. Results have been published concerning the development of bacterial cultures obtained from strains taken from amber tens of millions of years old (or from even older sediments). However, this has not been enough to convince most of the scientific community that they attribute

[35] For all the details of the discovery and analysis, see Meinesz (2008) and Fry (2000, Chap. 14).

these results to contamination with modern germs. That is, the researchers who claim to have successfully multiplied ancient microorganisms, would have done nothing more than multiply modern microorganisms that polluted the sample.[36] Currently, there is a lack of a procedure capable of guaranteeing the age of the microorganisms or genetic material under study. In this regard, Martin Hebsgaard and his collaborators (Hebsgaard et al. 2005) have proposed some criteria to be respected before announcing to the four winds an exceptional discovery on this subject. The most important, of course, is to make sure that you have decontaminated the sample from all possible modern pollutants a priori. Two other criteria deserve to be emphasized: the reproduction of the results in different laboratories and tests comparing fossil DNA with that of recent samples. The older the DNA, the higher the number of differences between the fossil sample and the control samples should be. The first of these seems essential to me, because science is based, rightly, on the ability to reproduce the results obtained.

The last criterion that decrees that fossil DNA (the genome of aged microbes) must be sufficiently different from that of modern samples, has aroused some criticism (Vreeland and Rosenzweig 2002). This is not the best place to delve into the subject. However, I would like to point out that, in my humble opinion, no one has yet been able to demonstrate the veracity of this assumption, even if, at first glance, it may seem logical.

Let us now return to our main argument, that terrestrial Life is actually originating from another planet. Nowadays, no certain evidence (accepted as such by a substantial part of the scientific community) has ever been able to prove this assumption, in an irrefutable way. Although various hypotheses have been put forward, a lot of evidence is still needed.

On the other hand, looking for the origin of Life elsewhere, because they are unable to explain the transition from non-living to living on Earth, is just a trick. All this highlights our inability to face and solve the problem. In my opinion, it hides the main problem, simply by setting the problem elsewhere.

In fact, the main question, established at the beginning of paragraph 1.2, remains valid: is it possible to conceive conditions that allow, one stage after another, to pass from a situation of absence of Life to one in which living beings can emerge spontaneously from a sterile world? In my opinion, Mario Ageno's hypothesis has the best basis to hope formulating a complete and correct theory on the subject, without the need to bother with extraterrestrial

[36] Among the partisans of the antiquity of such bodies, see Cano and Borucki (1995), Vreeland et al. (2000) and Powers et al. (2001). For more critical articles, see Graur and Pupko (2001), Hazen and Roedder (2001) and Maughan et al. (2002).

solutions. This, of course, does not mean that Life cannot have appeared elsewhere as well.

Ageno's proposition derives from a purely scientific conception: the transition from the non-living to the living is not the product of stages taking place more or less randomly. However, it constitutes the set of a series of processes, each of which represents the result of precise causes that determined the result. Each step implies and determines the following.

This means that once all the good conditions are gathered, Life, as we have defined it in the course of the chapter, emerges spontaneously. And this everywhere, in the Universe.

All this remains in the field of scientific research. At this stage, it is important to show that a scientific explanation of this process is not only desirable but also perfectly possible. And this is only the beginning... The fantastic story of biodiversity only begins!

1.6 Summary of This Chapter

When discussing living beings, we must start by defining the entities in question. How can we define a living organism? The Italian physicist Mario Ageno, during the 1990s, addressed this question by defining a living being as an open physical system (capable of exchanging materials and energy with the external environment), hosting coherent chemical processes (ordered in space and time) and equipped with a "program" that regulates its functioning (i.e., specifying where, what, and how occur). The simplest living organism meeting this definition is a cell, the minimal unit containing the information and all the mechanisms (adequately supplied) required to perform the metabolic processes essential for Life.

Ageno went further, attempting to lay the foundations of a theory explaining the transition from the non-living to the living. Today, every living being derives from another living being, but under appropriate conditions (likely those present during a specific period in Earth's), Life could have emerged from a sterile world. After the solidification of the Earth's mass and the end of meteor bombardment, around 3.8 billion years ago, our planet already featured vast expanses covered with water. The atmosphere at the time differed from today's, being composed mostly of water vapor, carbon dioxide (CO_2) and nitrogen gas (N_2), while gaseous oxygen (O_2) was absent.

During this period, terrestrial waters began to accumulate the so-called "primordial soup", a mixture of biological molecules (nucleotides, amino acids, lipid molecules, carbohydrates and/or their precursors) all synthesized abiotically, that is, in the absence of living beings. These molecules could have formed near hydrothermal vents or even in space. Their presence in the aquatic environment of the time is now widely accepted by the scientific community.

It is precisely from this particular soup, in a protected habitat (for example, a lagoon), near to the bottom but at a depth still reachable by sunlight, that Mario Ageno situates the subsequent stages of the process. The lipid molecules present in the soup would have formed pockets capable of fusing with one another yet also able to split without mixing their internal contents with the external environment. The lipid bilayer was indeed impermeable to most of the components of the primordial soup. It is within the system composed of ALL the pockets of the lagoon that the metabolic processes characterizing living beings gradually developed.

The key to this process lies in a simplified sequence of the photosynthesis mechanism, particularly the elements involved in cyclic phosphorylation. Such a mechanism would have allowed, through light energy captured by a pigment trapped in the lipid bilayer, the enrichment of the pockets' internal environment with H^+ ions coming from the outside. This enrichment would have been facilitated by a particular molecule capable of "swimming" through the bilayer of the sacs. The more acidic internal environment would have enabled not only the formation of phosphorylated molecules (equipped with phosphate groups able to release chemical energy for endergonic reactions) but also the production of active precursors of nucleic acids (DNA and RNA) and, consequently, the assembly of their strands.

With the production of nucleic acid molecules, Ageno further speculates on the creation of active precursors for peptide chains synthesis (macromolecules composed of amino acids) and the establishment of a code linking each amino acid to a triplet of nucleotide bases. This stage would have marked the realization of the process leading to living beings.

According to the Italian physicist's conception, the transition from the non-living to the living consists of a series of stages in which each step, from the beginning, determines and enables the next, starting from well-defined primary causes. If the conditions are properly met, the emergence of Life becomes an automatic, or nearly automatic, process.

Appendix

The 20 Fundamental Amino Acids

Each amino acid is encoded by a triplet of DNA nucleotides. After a protein sequence has been synthetized, however, the 20 fundamental amino acids can undergo various modifications: atoms of other elements can be added, certain bonds can be altered, and so on. Some particular organisms have even incorporated two additional "fundamental" amino acids, repurposing the coding previously used for other instructions. These amino acids are selenocysteine, which comprises a selenium atom (Se) and whose incorporation into proteins is otherwise very complex, and pyrrolysine. The latter is found exclusively in methanogenic archaea (archaea will be discussed in Sect. 2.2).

Quinones and Their Properties

Plastoquinone belongs to the quinone group. Their properties derive from the structure of alternating double bonds (Fig. 1.9).

On the left is the arrangement of the molecule BEFORE acquiring two electrons (e^-) and the two H^+ ions; on the right is the molecule AFTER gaining these electrons and ions. R1, R2, R3 and R4 represent four distinct chemical groups. This characteristic of quinones is shared by all molecules containing a double bond between two carbon atoms. Figure 1.10 is a simpler molecule (much simpler than other compounds obtained from Miller's abiotic experiments) that still exhibits this property.

On the left is the molecular arrangement BEFORE acquiring two electrons (e^-) and two H^+ ions; on the right is the arrangement AFTER electrons and ions acquisition.

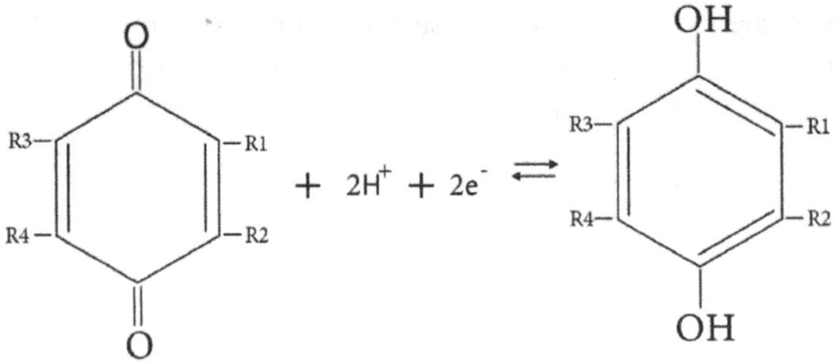

Fig. 1.9 Plastoquinone molecule, left before electron capture, right after electron capture. Redrawn from Ageno (1991, p. 303)

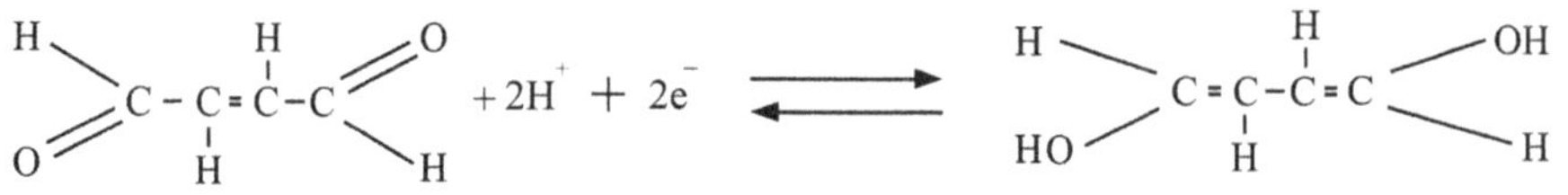

$$\mathrm{OHC{-}CH{=}CH{-}CHO} + 2H^{+} + 2e^{-} \rightleftharpoons \mathrm{HO(H)C{=}CH{-}CH{=}C(H)OH}$$

Fig. 1.10 Example of a very simple molecule but with double bonds allowing plastoquinone properties. Redrawn from Ageno (1991, p. 277)

References[37]

Ageno, M. 1986. *Le radici della biologia*. Milan: Feltrinelli.

Ageno, M. 1991. *Dal non vivente al vivente*. Rome-Naples: Theoria.

Ageno, M. 1992a. *La "macchina" batterica*. Rome: Lombardo Editore.

Ageno, M. 1992b. *Punti cardinali*. Milan: Sperling & Kupfer.

Cano, R.J., and M.K. Borucki. 1995. Revival and identification of bacteria spores in 25- to 40-million-year-old Dominican amber. *Science* 268: 1060–1064.

*Collective Work. 2006. De la vie sur Mars. *Sciences et Vie* 1060, January: 48.

*Collective Work. 2008. Où est née la Vie? *Dossier pour la Science*, July–September: 60.

*Collective Work. 2013a. Evolution—Les quatre premiers milliards d'années: de la cellule primitive à l'apparition des mammifères. *Geo savoir, hors-série*, February–March: 5.

*Collective Work. 2013b. Les origines de la vie. *Les dossiers de La Recherche*, February–March: 2.

Darwin, C. 2009. *On the origin of species*, 2nd ed. London: Natural History Museum.

Desbruyères, D., M. Segonzac, and M. Bright (eds.). 2006. *Handbook of deep-sea hydrothermal vent fauna*. Linz: Oberösterreichisches Landesmuseum Biologiezentrum.

*Forterre, P. 2007. *Microbes de l'enfer*. Paris: Belin pour la Science.

Forterre, P. 2010. Defining Life: The virus viewpoint. *Origin of Life and Evolution of the Biosphere* 40 (2): 151–160.

*Fry, I. 2000. *The emergency of Life on Earth—A historical and scientific overview*. New Brunswick (New Jersey): Rutgers University Press.

Geis, I. 1983. The DNA helix and how is it read. In *Scientific American*, vol. 249, ed. R.E. Dickerson, 97–112.

Graur, D., and T. Pupko. 2001. The Permian bacterium that isn't. *Molecular Biology and Evolution* 18 (6): 1146–1146.

*Gribaldo, S., M.-C. Maurel, and J. Vannier. 2007. *L'évolution—Les débuts de la vie*. Paris: Le Pommier.

[37] Works preceded by an asterisk can be read, more or less easily, by people with a non-professional skilling.

Hazen, R.M., and E. Roedder. 2001. How old are bacteria from the Permian age. *Nature* 411: 155–156.

Hebsgaard, M.B., M. Philips, and E. Willerslev. 2005. Geologically ancient DNA: Fact or artefact? *Trends in Microbiology* 13 (5): 212–220.

Jablonka, E., and M.J. Lamb. 2006. *Evolution in four dimensions—Genetic, epigenetic, behavioral, and symbolic variation in the history of Life*. Cambridge (Massachusetts): The MIT Press.

*Lane, N. 2010. *Life ascending—The ten great inventions of evolution.* London: Profile Books.

*Lane, N. 2015. *The vital question—Why is Life the way it is?* London: Profile Books.

Lecointre, G., ed. 2009. *Guide critique de l'évolution.* Paris: Belin.

Maughan, H., C.W. Birky Jr., L.W. Nicholson, W.D. Rosenzweig, and R.H. Vreeland. 2002. The paradox of the "ancient" bacterium which contains "modern" protein-coding genes. *Molecular Biology and Evolution* 19 (9): 1637–1639.

*Maurel, M.-C. 2003. *La naissance de la Vie—De l'évolution prébiotique à l'évolution biologique*, 3rd ed. Paris: Dunod.

*Maynard Smith, J., and E. Szathmáry. 1999. *The origins of Life. From the birth of Life of the origins of language*. Oxford University Press.

*Meinesz, A. 2008. *Comment la vie a commence—Les trois genèses du vivant*. Paris: Belin.

*Mieli, E., A.M.F. Valli, and C. Maccone. 2025. *The living galaxy—Winners and losers in the Milky Way*. Cham, Switzerland: Springer Nature.

Miller, S. 1953. A production of amino acids under possible primitive Earth conditions. *Science* 117: 528–529.

Powers, D.W., R.H. Vreeland, and W.D. Rosenzweig. 2001. How old are bacteria from the Permian age?—Reply. *Nature* 411: 155–156.

Powner, M.W., B. Gerland, and J.D. Sutherland. 2009. Synthesis of activated pyrimidine ribonucleotides in prebiotically plausible conditions. *Nature* 459: 239–242.

Purves, W.K., D. Sadava, G.H. Orians, and H.C. Heller. 2003. *The science of the biology*, 7th ed. Sinauer Associates and WH Freeman.

Rasmussen, B. 2000. Filamentous microfossils in a 3,235 million-year-old volcanogenic massive sulphide deposit. *Nature* 405: 676–679.

Ritson, D., and J.D. Sutherland. 2012. Prebiotic synthesis of simple sugars by photoredox systems chemistry. *Nature Chemistry* 4: 489–899.

Rosing, M.T., and R. Frei. 2004. U-rich Archaean sea-floor sediments from Greenland—Indications of > 3700 Ma oxygenic photosynthesis. *Earth and Planetary Science Letters* 217: 237–244.

Schopf, J.M., and B.M. Parcker. 1987. Early Archean (3.3 billion to 3.5 billion-year old) microfossil from Warrawoona Group, Australia. *Science* 237: 70–73.

Shen, Y., R. Buick, and D.E. Canfield. 2001. Isotopic evidence for microbial sulphate reduction in the early Archaean era. *Nature* 410: 77–81.

Sojo, V., B. Herschy, A. Whincher, E. Camprubí, and N. Lane. 2016. The origin of Life in alkaline hydrothermal vents. *Astrobiology* 16: 181–200.

Taylor, T.N., E.L. Taylor, and M. Krings. 2009. *Paleobotany: The biology and evolution of fossil plants*, 2nd ed. Amsterdam: Academic Press.

Tice, M.M., and D.R. Lowe. 2004. Photosynthetic microbial mats in the 3,416 Myr-old-ocean. *Nature* 431: 549–552.

Vreeland, R.H., and W.D. Rosenzweig. 2002. The question of uniqueness of ancient bacteria. *Journal of Industrial Microbiology and Biotechnology* 28: 32–41.

Vreeland, R.H., W.D. Rosenzweig, and D.W. Powers. 2000. Isolation of a 250 million-year-old halotolerant bacterium from a primary salt crystal. *Nature* 407: 897–900.

Wacey, D., M.R. Kilburn, M. Saunders, J. Cliff, and M.D. Brasier. 2011. Microfossils of sulphur-metabolizing cells in 3.4-billion-year-old rocks of Western Australia. *Nature Geoscience* 4: 698–702.

Wilson, E.O. 1975. *Sociobiology—The new synthesis.* Cambridge (Massachusetts): Harvard University Press.

2

The First Living Beings

2.1 The Oldest Traces of Terrestrial Living Organisms

In the previous chapter, we explored how Life could have originated on our planet. I attempted to explain, as succinctly as possible, the development of the main events, following the framework proposed by Mario Ageno. In my view, this represents the most accurate and comprehensive set of ideas for constructing a preliminary theory on the subject.

After examining the conditions that allowed Life to emerge from a completely sterile environment and summarizing the main stages of the process, we are ready to address several important questions. When did Life appear on Earth? What did the first living organisms look like?

To answer these questions, it is first necessary to identify fossil traces of the earliest living beings, however tiny and difficult to recognize they may be. Since the first organisms were probably unicellular, the challenge essentially lies in detecting traces of the earliest cells in extremely ancient sediments.

Several criteria have been proposed for this purpose, which full into two categories: chemical and morphological.

Chemical criteria are considered among the most robust, though they are not always conclusive. For instance, it involves recognizing, within very ancient (et potentially heavily altered) sediments, the imprints of chemical processes resulting from the degradation of organic molecules (processes produced by living organisms). Calvin (1969) and his collaborators were able to detect, through traces of chemicals synthesized by living organisms, the presence of fossils in 3.3 billion years old (3.3 Ga) layers in Swaziland (Africa).

A. M. F. Valli, *The Three Domains of Life*, Copernicus Books,
https://doi.org/10.1007/978-3-032-14802-5_2

Building with these findings, the team expanded its investigation to other rocks, again searching for fossil traces.

While the results of the 1960s and 70s initially represented a major success, it later became clear that chemical traces of hydrocarbons, such as those found by Calvin, could also be produced through abiotic processes.[1] These discoveries introduced a degree of uncertainty regarding earliest results.

The presence of kerogen[2] has often been citated as reliable evidence of the biological origin of the carbon in sediments, but it is not always straightforward to determinate whether its origin is biological, abiotic, or mixed. Another indicator is the relative abundance of stable carbon isotopes.

What are these isotopes? They all belong to the element carbon (meaning they react chemically like carbon) but they differ in one kay aspect: their mass.

Every atom of a given element has a fixed number of electrons and protons. Hydrogen, for example, has only one electron (negatively charged) and one proton (positively charged), whereas oxygen has eight electrons and eight protons, totalling 16 charges, balanced between positive and negative.

The number of protons and electrons determines an atom's identity and properties. Since the atom is neutral, both electric charges balance out. This means that each atom has the same number of electrons and protons. However, the two particles differ also in mass. The electron weight is negligible compared to that of the proton.[3]

Atoms also contain neutrons, which are electrically neutral but have a mass roughly equal to that of the proton. All atoms of an element (except hydrogen) have at least two or more neutrons, which increase the mass but not the properties of the element.

Additionally, the number of neutrons in atom of the same element can vary. If we consider two or more atoms of the same species, they will have the equal number of electrons and protons (otherwise they would belong to different elements), but they may differ in neutron number. For example, most carbon atoms have six neutrons (we call them ^{12}C, where 12 means "six protons and six neutrons"), while some have seven neutrons (we call them ^{13}C, because they combine six protons with seven neutrons; $6 + 7 = 13$). ^{12}C and ^{13}C are two stable carbon isotopes.[4]

[1] For further information on the chemical and morphological methodologies which can identify traces of past micro-organisms, the interested reader may consult the book by Ageno (1991, pp. 106–110).

[2] Kerogen is an intermediate compound in the transformation of organic matter into fossil fuel.

[3] See also footnote 11 of Chap. 1.

[4] Another carbon isotope has become very famous, thanks to its use for the dating of prehistoric samples: it is ^{14}C. But beware, this is not a stable isotope; it is radioactive. In fact, it disintegrates over time and changes its nature. It is therefore not suitable for the purposes we are interested in.

What is the difference between ^{12}C and ^{13}C? Both are carbon atoms, but ^{13}C is a little heavier, because it has an additional neutron. This means that in the same reactions, one isotope reacts slightly faster or slower than the other, due to the difference in mass. This has an important consequence: if the initial $^{12}C/^{13}C$ ratio (X1) is known, the ratio (X2) after the reaction can provide a characteristic signature of the process. Detailed studies have established typical $^{12}C/^{13}C$ ratios for biological processes, allowing scientists to use isotope ratios as markers for distinguishing biological carbon from non-biological sources.

However, this method has limitations, as ambiguous values may occur. Under certain conditions, abiotic processes (such as those in hydrothermal fluids) can produce carbonate components with $^{12}C/^{13}C$ ratios similar to those of biological origin.

Morphological criteria differ entirely: they involve examining the shapes of fossil imprints in sediments and comparing them with those of known organisms. In the past, this method was considered unreliable, because recent microorganisms often have very simple external morphologies that could resemble natural traces. Nevertheless, improved techniques have increased its reliability.

For instance, in 1977, a team of researchers (Muir et al. 1977) conducted detailed analyses of roughly spherical impressions found in African strata dated to approximately 3.35 Ga. The selection of traces for analysis followed strict criteria, concerning very precise morphological and physical characteristics and contemporaneity within the sediments, while excluding ambiguous structures with fractures or recrystallizations.

Once the traces were appropriately selected, they were closely compared with a sample of more recent, authenticated microfossils. The strong statistical agreement between the two samples confirmed the biological origin of the oldest fossils.

Despite this success, it must be acknowledged that obtaining unambiguous and usable samples in not always possible. The older the sediment layers, the greater the difficulty. Therefore, morphological analyses cannot always be performed successfully.

Another morphological criterion for identifying ancient life forms is stromatolites, whose distinctive structure is reliably characteristic. But what are stromatolites? The term does not refer to an organism itself, but to a carbonate sedimentary structure (a deposit of carbonate sediments with a specific shape and structure). Crucially, the formation of stromatolites is driven by biological factors, making them valuable indicators of early life.

In section, this structure is easily recognizable, because it shows an alternation of very thin layers (laminations) and corrugated layers (Fig. 2.1). Consider a *mille-feuille* cake and cut a slice: the section of a stromatolite resembles the piece obtained. Only the layers will not be straight but wavy.

Although stromatolites are not comparable to bones or shells, they are true fossils because they testify to the presence of living beings. In the case of stromatolites, they are entire communities of different organisms, with one dominant species and others less abundant, that allowed the construction of the structure.

Photosynthetic organisms (= capable of carrying out photosynthesis), whether cyanobacteria[5] or green algae, form the bulk of the biological mass, but they are also accompanied by other creatures of different sizes, from microscopic microorganisms to small animals, which in same case can reach the size of a small crustacean (a few fractions of a millimetre). These organisms take advantage of the activity of photosynthesizers and the shelter offered by layers of deposited sediment as the structure grows.

Now let us see how a stromatolitic structure is formed. Photosynthetic organisms settle at the bottom of a body of water (fresh or salt), on hard

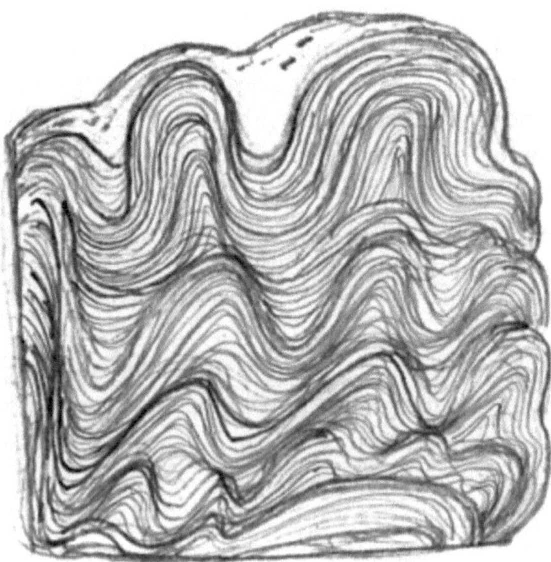

Fig. 2.1 Section of a stromatolitic structure. The lamination of the layers is characteristic of this type of fossil

[5] Cyanobacteria are special bacteria, capable of photosynthesis comparable to that carried out by green plants, with production of oxygen.

and rigid surface, not too deep because they need light. There, they form a biological mat that encrusts the substrate and shapes its surface: this is how the undulations they visible in section are formed.

Whether the secreted mat traps the deposited sediments like glue, or whether the secretions of certain microorganisms favour calcite precipitation (or both phenomena act simultaneously), eventually a layer of debris will cover the community. The photosynthetic organisms, which need light, then settle above the sediments to continue their activity. The cycle repeats until the organisms are once again covered by another layer of sediment. Stratification continues, resulting a series of juxtaposed layers observable in the section.

Over time, depending on factors such as sedimentation rate, subsidence,[6] or the variations in water layer, a structure of varying size and complexity forms.[7] Although external morphology is influenced by various factors, stromatolitic construction always show, in section, a characteristic and easily recognizable alternation of laminated and undulating layers.

A very old type of structure is among the oldest known fossils, but stromatolites are found throughout geological history, from the oldest to the present day.[8] They are very common between 2.5 and 0.5 Gy, peaking around 1.2 Gy. Afterwards, stromatolites undergo a slow decline, despite two additional peaks during the Palaeozoic, the first around 500 Mya, the second around 350 My,[9] until they reach current abundance levels (Taylor et al. 2009, p. 67, Fig. 2.35).

Identifying such structures, especially those built in shallow environments far from hydrothermal vents or abundant terrigenous sediments, is considered one of the best criteria for recognizing life form characteristic of the most ancient geological periods. Of course, evidence becomes even more convincing when other criteria, such as chemical ones, are added to the context.

These are the tools available to palaeontologists to identify fossils in the oldest sedimentary rocks. Armed with these tools, an increasing number of researchers have investigated the oldest sedimentary formations to find traces of Life.

---

6 Subsidence is a slow lowering of the soil, which can cause progressive deposition of sediment.

7 There is a relatively large body of scientific literature on stromatolites. Here are some selected articles, which can be read with profit, in English (Hoffman 1967; Freytet and Verrecchia 1998; Freytet 2000; Awramik 2006) as well as in French (Bertrand-Sarfati et al. 1966; Wattinne et al. 2003).

8 Among the current stromatolites, the best known in marine environment are those of Shark Bay, Australia. In the lake environment, however, we must mention those of Cuatro Ciénégas, in Mexico.

9 Remember that "My" means "millions of years" while "Gy" means "billions of years".

I have already mentioned that fragment of rock that dates back a long time (paragraph 1.5); it is the ALH84001 meteorite, which fell in the Antarctic and came from Mars. However, it is an extraterrestrial rock! I have already discussed all the problems related to the nature of its content. Here, let us limit to the rocks that have formed on our planet.

Among the oldest terrestrial sedimentary rocks are those discovered in Greenland, on the island of Akilia, south of Nuuk, and those of the Isua Formation in western Greenland. The Akilia strata are older than 3.850 Gy, and the Isua Belt dates to between 3.7 and 3.8 Gy.

In Africa, the Onverwacht geological series spans the region between South Africa and Swaziland, with rock ages ranging from 3.96 to 2.5 Gy; the older layers rival Greenlandic rocks in age. Also in Africa, the Barberton Greenston Belt, in the mountains around the city of the same name, dates to about 3.5–3.2 Gy.

Very old sedimentary rocks are also found in Australia, slightly younger than Greenlandic ones but still venerable in age. In the northwestern Australia, the "Pilbara Supergroup" includes several sub-units ranging from more than 3.52 to about 3.0 Gy, including the Warrawoona Group (3.52–3.43 Gy) and the Kelly Group (3.43–3.31 Gy).

All of these locations have been intensely studied, and their sediments continue to be analysed to discover traces of Life (or to challenge previously reported results). Currently, after extensive discussions, the scientific community generally recognize the traces found in 3.5 Gy Australian and African sediments (or, possibly, younger) as authentic fossils.[10] Opinions remain divided regarding older strata (~ 3.8 Gy), with some researchers considering the fossil traces authentic, while others believe that metamorphism and tectonic processes erased any evidence of life (if there ever were any), leaving only artifacts of abiotic origin.[11] Publications still report biological traces from 3.8 Gy rocks (Rosing and Frei 2004),[12] but further research is need.

---

[10] A little bibliography, presented in chronological order, according to the date of publication: Schopf and Parcker (1987), Rasmussen (2000), Shen et al. (2001), Brasier et al. (2002), Tice and Lowe (2004), Allwood et al. (2006, 2007) and Wacey et al. (2011).

[11] Among those who consider the fossils found in rocks of 3.8 Gy age to be authentic, see Mojzsis et al. (1997) and Holland (1997). For sceptics, Fedo and Whitehouse (2002) and Van Zuilen et al. (2002). The interested reader may also wish to consult Chap. 13 of the book by Fry (2000) and the volume by Knoll (2004). Recently, the article Nutman et al. (2016) reports evidence of authentic stromatolites in the Iusa Formation, in Greenland, 3.7 Gy aged, and Dodd et al. (2017) claims to have recognized even older fossils in the Nuvvuagittuq Formation (Québec, Canada): the debate is relaunched!

[12] Work already cited in footnote 34 of Chap. 1. See also previous footnote.

2.2 The Three Domains of Living Things and Biodiversity During the Archean

We have just explored the oldest known organic remains of our planet. Those generally accepted by the majority of the scientific community date to approximately 3.5 Gy. What kind of fossils have been found in these ancient sediments?

In the Australian rocks (Warrawoona Group), researchers discovered stromatolitic structures formed in shallow aquatic environments, as well as traces of microorganisms, either isolated, grouped in small units, or arranged in filaments. Similar traces have also been found in African sediments (Barberton Greenston Belt). The most characteristic organisms have been identified as cyanobacteria based on their anatomical characteristics. These bacteria can perform photosynthesis in both in oxygenated and anoxic conditions. They are also involved in the construction of modern stromatolitic formations.

Fossil evidence thus indicates that these organisms (or very close related forms) were already part of the microbial fauna of the time.

But what exactly is a cyanobacterium? It is a single-celled organism, without a nucleus (therefore a real bacterium), but capable of the same green plants' performance. These microbes, in fact, are able to perform photosynthesis with emission of oxygen (a metabolic product of the process[13]), as plants do.

Other fossil and chemical evidence point the presence of additional organisms. Scientists have found fossil microbes with sulphate-reducing metabolism. Various clues suggest the existence of very ancient hydrothermal springs. Today, most organisms with this metabolism belong to a particular group once considered a simple subset of bacteria, now recognized as a distinct super-kingdom (or domain): the archaea.[14]

[13] It should be noted that not all photosynthesis processes produce oxygen. This gas is generated only when water supplies the electrons lost from the chlorophyll of the photosynthetic apparatus. This is the case of green plants and some micro-organisms, including cyanobacteria (this is the process we have already discussed, albeit in a succinct way, in paragraph 1.4). When it comes to another substance to provide the electrons, you no longer get oxygen, but another "waste" that depends on the nature of the electron donor. For example, the members of the Halobacteria class are microbes that like environments with very high salt concentration. These organisms are capable of photosynthesis, but do not release any oxygen molecules. In reality, these microorganisms are not real bacteria, but of archaea. To know better this type of organisms, continue reading the text.

[14] The recognition of this group and its importance is due to Professor Carl R. Woese (Woese and Fox 1977; Woese et al. 1990). For all those who want to know more about these microorganisms, I recommend reading a very pleasant book, in French, written by Forterre (2007). Among the oldest fossil traces of this group, one should not forget the inclusions of methane, considered to be of biological origin, which are about 3.5 Gy (Ueno et al. 2006).

So, what are these organisms, and how do they differ from ordinary bacteria? Archaea are single-celled prokaryotes; their cells have no nucleus and their DNA is ring-shaped and located directly in the cytoplasm (the internal environment of the cell). Yet, they display biochemical and structural characteristics that phylogenetically[15] distinguish them both from bacteria and eukaryotes. Eukaryotes include all organisms whose cells contain a double-membrane nucleus.[16] The eukaryote group also encompasses various unicellular organisms, such as amoebas and paramecia, as well as all multicellular organisms: animals, fungi and plants.

Although, initially only a few archaeal species were recognized, today we know that this group is much more diverse. Most inhabit extreme environments, considered "hellish" in our eyes. For example, some archaea, called "hyperthermophiles", thrive at temperatures near boiling (hence the name), such as in deep-sea hydrothermal vents, volcanic ponds, or hundreds of meters underground high pressure and temperature.

Others prefer highly saline environments, such as the Dead Sea. Among the archaea, we find the smallest known organism, *Nanoarchaeum equitans*, which lives in underwater hot springs off the coast of Iceland, in complete absence of oxygen.[17]

However, not all archaea require such extreme conditions. The first recognized archaea were methanogenic microorganisms, which generate energy by reacting carbon dioxide (CO_2) with hydrogen molecules (H_2).

The process, which sees four hydrogen molecules react with one molecule of CO_2, produces two molecules of water (H_2O) plus one molecule of methane (CH_4). These microbes thrive in anaerobic (oxygen-free) environments, like lake basins, the deep sea, or even… human intestine! In fact, inside our body, microorganisms (bacteria but not only) are numerous. Some

---

15 Phylogeny is the science that deals with relationships between living beings, so as to understand who is most related to whom. The results are generally given in the form of a "tree" as shown in Fig. 2.3, paragraph 2.3.

16 Some microorganisms belonging to Planctomycetota (phylum of aquatic bacteria that reproduce by budding, typical of brackish soils, but also found in fresh or salt water) possess a membrane called introcytoplasmic (ICM), differently developed, depending on the organism in question, which encloses all the cellular DNA and also the ribosomal material. Furthermore, the genome of *Gemmata obscuriglobus* is wrapped in a double membrane which is itself contained in the ICM (Lindsay et al. 2001). Although no homology has yet been revealed, the similarity with a eukaryotic structure is impressive. However, the nuclear membrane is only one of the characteristics of eukaryotes. Others, such as the presence of particular organelles (mitochondria and chloroplasts), contribute to the definition of this domain of the living. The next chapter will be devoted to them; in that place, I will evoke the evolutionary importance of the eukaryotic organization of the cell.

17 Forterre's book (2007, Annex 3) presents the list of known archaea. As for *Nanoarchaeum equitans*, this microbe has a spherical shape with a diameter of about 400 nm (nanometres). The "nm" corresponds to a billionth of a metre, that is, it takes a billion nanometres to obtain a length equal to one metre!

of these, such as *Escherichia coli*, are so efficient at consuming the oxygen, that they allow other microbes, for those this gas is a toxic, to thrive without any problem!

According to Professor Woese, the first to discover the reality of archaea, the phylogeny of living beings is not based on the dichotomy "prokaryotes/eukaryotes"[18] but, rather, on a trichotomy including three domains: that of archaea (Archaea), that of true bacteria (Bacteria) and that of eukaryotes (Eukaryota).

Eukaryotes differ from bacteria and archaea in their cell type, which has a double-membraned nucleus. Prokaryotes (bacteria and archaea) have no nucleus and their cell is bounded by a membrane plus a rigid wall. The wall gives each species its characteristic morphology (spherical form, a rod, and so on). Eukaryotes, on the other hand, only possess the membrane. However, they have organelles (e.g., mitochondria and chloroplasts), which are unknown among prokaryotes. Into the organelles occur particular biological processes. For example, chloroplasts are only present in photosynthesizing organisms (green plants and single-celled algae). If the body loses them, it also loses any photosynthesis ability. Instead, mitochondria produce the biological energy of the cell. In the next chapter we will return to mitochondria and chloroplasts. Remember them!

Bacteria and archaea lack organelles; photosynthesis (in the case of cyanobacteria) or energy production are performed without such structures.

Despite having the same prokaryotic cell organization (cells without a nucleus), bacteria and archaea are very different. Let us start by dealing with the cell membrane. Bacteria have a double lipid wall, like the one we already seen in the previous chapter, when I discussed Mario Ageno's hypothesis. Hydrophobic tails (those that do not like water) are formed by linear fatty acids. However, in the case of archaea, the chemical constitution of the same structures is different, because some chains have short branches. In addition, in their membranes, they use types of binding that are unknown in analogous bacterial or eukaryotic structures.[19]

---

[18] In any case, same recent data are better suited to a dichotomy Bacteria/Archaea (Williams et al. 2013).

[19] In the cellular membrane of the archaea, the hydrophobic tails of the elements that constitute the double lipid state are linked to the hydrophilic heads by means of "ether" bonds (of type R-O-R1, where O is oxygen while R and R1 are two atoms of carbon or another element), more resistant to high temperatures. In contrast, bacterial and eukaryotic wall use "ester" bonds, like the one shown below, which are less resistant under the same conditions.

$$R - \overset{\overset{\displaystyle O}{\|}}{C} - O - R_1$$

It seems that these characteristics allow Archaean cell membranes to maintain ion impermeability even at extreme conditions (e.g., temperatures close to that of boiling water). In fact, some archaea thrive in exceptional temperature and pressure conditions.

But not everyone lives in such rigorous environments: methanogenic archaea inhabit different environments, but all characterized by the lack of free oxygen. Their cells possess a cell membrane of the same type as that of hyperthermophilic archaea. How come? It is probably simply an evolutionary legacy. The first archaea possessed this membrane and transmitted it to their descendants. In addition, archaea lack muramic acid in the cell wall. This molecule is present in all known bacteria.[20] In this too, we must think of a phenomenon of evolutionary inheritance.

Therefore, although bacteria and archaea both possess the same type of cell (prokaryote), there are important structural differences between them. Another, even more important, was highlighted by Professor Woese. I will talk about it, however, later, because it concerns the differences between the three groups (bacteria, archaea and eukaryotes) and not only prokaryotes.

Let us now consider the similarities between eukaryotes and each of the two prokaryotic groups. Apart from the common characteristics of the three groups (discussed in the next chapter), cellular organisms with nuclei have in common with true bacteria the structure of the lipid double-walled membrane different from that of archaea. But are there any characteristics in common between the latter and eukaryotes? Characters that the two groups do not share with bacteria?

The answer is yes! The structure of the RNA polymerase (the enzyme complex responsible for the synthesis of RNA from a DNA template) of archaea is more complex and eukaryote-like than that of bacteria. Bacteria have only one type, a complex composed by four sub-units (four protein chains synthesized and assembled to obtain one functional enzyme). However, eukaryotes possess three different RNA polymerases, composed of a number of sub-units varying between 12 and 14.

Archaea have only one RNA polymerase, like bacteria, but that is composed of 13 sub-units (what is unusual for prokaryotes). From this point of view, the RNA polymerase of archaea is more similar to that of eukaryotes than to that of bacteria, despite the prokaryotic organization of the cell.

The similarities are not limited only to transcription (synthesis of RNA from a DNA template). Certain mechanisms and proteins involved in the

---

See Forterre (2007, pp. 118, 222–223).

[20] Eukaryotes do not have a cell wall. Therefore, their isolated cells are generally less rigid than those of prokaryotes.

replication (the process of DNA synthesis) and translation (protein production from a RNA strand: it takes place in ribosomes) of archaea have characteristics in common with those of eukaryotes. Archaea are therefore very special organisms!

Another notable feature of archaea, different from both bacteria and eukaryotes, concerns ribosomal RNA (rRNA). Ribosomes[21] (already introduced in the previous chapter), are indispensable to Life, because they are the site of protein synthesis, starting from an RNA molecule that determines the sequence of amino acids that must be synthesized. This process is called "translation": an RNA strand is translated into a protein sequence. Without ribosomes, proteins cannot be obtained, and cellular function ceases.

If we now consider the rRNA sequence of the smallest sub-unit, we find an unexpected result. Archaea sequences differ as much from bacteria as bacteria do from eukaryotes! So, despite their prokaryotic cell organisation, archaea are thus unique organisms, phylogenetically distinct from both bacteria and eukaryotes (Woese and Fox 1977; Woese et al. 1990). Archaea and bacteria are both prokaryotes, but prokaryotes of a very different kind!

With all these considerations in mind, let us move on to examine what we know about biodiversity during the earliest periods of our planet.

Let us take a look at the Archean, the oldest eon[22] since the solidification of the Earth's crust (between 4.2 and 4 Gy). The Archean, spanning from 4 to 2.5 Gy, covers a time interval of 1.5 Gy. This is one of the longest time intervals recognized on our timescale: an immense duration! All fossils know by the general public span a time interval of about 600 My; less than half Archean duration!

At one time, our knowledge of fossils belonging to this eon was extremely limited. Nowadays, however, we can try to establish a somewhat more complete picture of the main forms of life of the time.

---

[21] A ribosome is made up of two sets; one larger and one smaller. Each set (sub-unit) comprises an RNA molecule (the rRNA) and a peptide chain. The latter is a sequence of amino acids, like a protein. Unlike proteins, the latter comprises on average only 300 monomers, although it may be longer and more complex in certain cases (for example when it is composed of several sub-units). A simple peptide sequence can be even shorter. When it includes only some amino acid it is also called "oligopeptide" or "oligopeptide sequence".

[22] Aeons are among the longest divisions of the Earth's chronological time. They have an exceptional duration, several hundred million years. Eons can be divided into eras, which in turn are composed of periods and so on, with ever shorter divisions, to meet the needs and chronological knowledge. The international time scale, in several languages, is available on the website of the International Commission on Stratigraphy.

During Archean, Life was restricted to aquatic environment, because, with the modern atmosphere still lacking, the deadly radiation from space (especially ultraviolet rays) would destroy all unprotected organisms. However, in water, living beings are sheltered.

On the seafloor, in shallow depths, photosynthetic microorganisms were dominant. Some formed biological mats or stromatolites (I remember they are among the oldest known fossils). Cyanobacteria, but also organisms that do not produce gaseous oxygen (capable of photosynthesis without producing O_2), occupied such ecological niches.

Stromatolitic mats and formations shared the aquatic environment with other types of organisms. These used various and differentiated strategies: filamentous microorganisms (bacteria?), more or less thick or tubular in shape, colonies of unicellular beings, spherical microbes, cells of different shapes and sizes. In 3.2 Gy layers, some spherical fossils are almost 300 μm in diameter (Javaux et al. 2010) (a μm, micrometre, is the millionth part of a meter; that is, there are one million micrometres in a meter). In the Australian Warrawoona Group layers (older than 3.35 Gy), four different types of microbial filaments have been recognized: each received a different generic name (Taylor et al. 2009, Chap. 2). Of course, we also have to consider single, isolated cells.

Methanogenic microbes were probably common, given that oxygen was scarce at the time while CO_2 was relatively abundant. In the depths, near the hydrothermal springs, where the temperature could exceed 100°, autotrophic chemosynthetic organisms thrived, especially those able to live by exploiting the sulphate-reduction process. Finally, the water column harboured cyanobacteria and other floating organisms.

To summarize, the Archean specialists have identified something like 48 different stromatolite sites (with structures having different shapes and sizes) and no less than 40 different morphotypes of presumed microfossils (Fig. 2.2), found in the most varied paleoenvironments, from shallow depths to hydrothermal vents (Nisbet 2000; Schopf et al. 2007). This list is destined to grow as the years go by.

2.3 LUCA and His Descendants

In the previous section, we saw that by 3.5 Gy, the Earth was already inhabited by organisms considered similar to cyanobacteria and sulphate-reducing microbes. The latter mainly belong to the archaea group and thrive in high-temperatures environments, such as underwater hydrothermal vents.

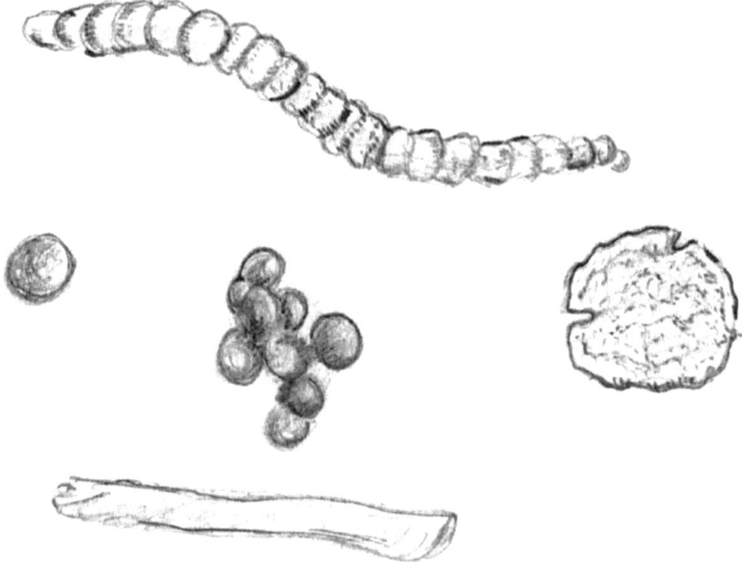

Fig. 2.2 Here are some reconstructions of the microfossils found in the layers of the Archean (the specimens are not in scale with each other). Above, a filamentous colony of *Primaevifilum amoenum* from Apex Chert (Pilbara Supergroup, north-western Australia) of ~ 3465 My; total length, about 30 μm. In the middle, on the left, a coccoid microbe from the Monte Cristo Formation (South Africa) of ~ 2600 My; the diameter is about 3 μm. In the middle, on the centre, a colony of spherical cells from the Warrawoona Group (Pilbara Supergroup, Northwest Australia) of ~ 3430 My; the diameter of a single cell measures about 8 μm. In the middle, on the right, a spherical microorganism from the Moodies Group (South Africa) of ~ 3200 My; the diameter of the structure is almost 300 μm. Below, a tubular case of *Siphonophycus transvaalense* from the Gamohaan Formation (South Africa) of ~ 2516 My; total length about 150 μm

We have also noted that Professor Carl Woese considers true bacteria (including cyanobacteria) to be phylogenetically very distinct archaea (Woese and Fox 1977; Woese et al. 1990).[23] This means that the common ancestor of both groups lived in an even older epoch on the chronological time scale.

Consequently, the fossils found in the 3.5 Gy layers do not correspond to those of the oldest terrestrial organisms, but only to the oldest known ones. Others must have preceded them, living in even older times.

But there is another possibility: why don't we simply admit that bacteria and archaea appeared independently of each other, at the moment of the

[23] I remember that Carl Woese showed not only that archaea batteries are phylogenetically distinct but that the two groups differ strongly, and in equal way, even from eukaryotes.

transition from the non-living to the living? And that is that Life has been produced more than once on our planet.

If we examine again the of Mario Ageno's proposition, there is nothing to prevent us from thinking that the transition from the non-living to the living could have taken place in several lagoons at the same time and produced different living beings. Ageno's hypothesis does not prevent such a possibility.

None of the proposals put forward about the origin of Life prevents it from not having appeared several times. For example, the transition from the non-living to the living could have occurred in different hydrothermal vents and the descendants could have evolved along very different evolutionary lines from each other. Alternatively, living organisms could have arrived from various planets where they could have made their appearances in different ways from each other.

So, why not consider that the ancestors of cyanobacteria appeared in a lagoon, as indicated by Mario Ageno, and the archaea (or their precursors) in a deep hydrothermal spring? There are researchers who think that the lifestyle of hyperthermophiles may represent the oldest of all living beings on our planet.

However, if we examine the characteristics common to ALL modern living organisms, eukaryotes, bacteria or archaea, a very different picture emerges. Let us begin by considering the characteristics of all organisms.

The genetic heritage of all living beings belonging to the three domains of Woese is contained in their DNA[24] and this molecule always consists of the same four nucleotides (all use the deoxyribose molecule), presented in the previous chapter. All other possible nucleotides with different bases are absent from DNA molecules. Furthermore, the genetic code is identical in all living organisms. It is based on a codon (genetic sequence that codes for an amino acid) of three nucleotides (a combination of three out of four possible nucleotides, i.e. $4^3 = 64$). Since there are only 20 standard amino acids (those attached to tRNA; the others are obtained by modifying those already integrated into the peptide sequence[25]) this means that some are coded for more than one codon.

DNA forms the double helix, which has become famous in biology books in all living beings. Recall that this structure is made possible by the presence

[24] Only a few viruses, including the famous AIDS virus, use RNA to store genetic information.

[25] At least two exceptions to this are known. These are selenocysteine and pyrrolysine (see paragraph 1.2). However, these two cases can be explained as evolutionary achievements obtained in two evolutionary lines of different organisms, sacrificing a triplet used differently (at first, it was a terminating codon), to supplement a particular amino acid otherwise obtained by modifying one of the fundamentals already integrated in the peptide sequence.

of deoxyribose in its skeleton; RNA, which has simple ribose in its place, is incapable of forming stable double helices!

If we move on to consider RNA, we can see that this molecule always has the same characteristics in all living beings. It is always made up of the same units[26] and is produced by RNA polymerase from a strand of DNA.

Proteins, which express the properties and capabilities of the cell, are manufactured on ribosomes (always present and indispensable in ALL organisms—bacteria, eukaryotes, archaea). All ribosomes are made up of two subunits, one large and one small, although evolution has produced differences (these are currently used to reconstruct the phylogenetic relationships of all living beings[27]). Messenger RNA (mRNA) reaches the ribosomes with information from DNA and the protein is produced, one amino acid after another. Amino acids are transported by tRNA, characterized by their anticodons, the complementary sequences to the codons (triplets) present on the mRNA.

The list of common characteristics could go on, but the above is enough to prove that ALL current eukaryotes, bacteria and archaea derive from the SAME ancestor. In fact, an ancestor common to all living things would be 10^{3489} more likely than the best scheme based on a multiple origin[28]!

So, who is this common ancestor? The community has given it a name; LUCA (Penny and Poole 1999; Forterre et al. 2005), which is the acronym for "Last Universal Common Ancestor", the most recent ancestor common to all living beings, or more familiarly, the last (the most recent) common ancestor!

Note that the existence of this common ancestor does not preclude the possibility that Life could have emerged more than once on our planet. It simply establishes that all modern organisms descend from a common ancestor and, if we go far enough back in time, we can hope to meet it.

But when would the hypothetical common ancestor have lived? Since the oldest known fossils date to about 3.5 Gy, and that, among these organisms, bacteria have already differentiated from archaea, LUCA must have lived between this date and 4–3.8 Gy, when the great meteorite bombardment on Earth ended (the date from which Life on Earth could have begun to develop).

---

26 See Fig. 1.7, in Chap. 1.

27 This is exactly what Woese did and which allowed him to highlight the dominance of the archaea. See the two preceding paragraphs.

28 This information and the characteristics mentioned above were obtained from the English site of Wikipedia dedicated to LUCA, acronym that means "the Last Universal Common Ancestor" (the most recent ancestor common to all living beings), of which we are going to talk.

More likely, LUCA lived very close in time to the first bacterial and archaeal cells of bacteria and archaea. In fact, the portrait established for LUCA is that of an already complex organism. It had to possess the set of characteristics common to the three domains. LUCA was a cellular organism, with a few thousand proteins capable of performing its vital functions (other researchers suggest it required even fewer: it would have needed only 500–600 genes; Koonin 2003). Its energy source for manufacturing phosphate-containing molecules such as ATP, and its carbon source, would have been glucose.

Was it a DNA-based organism or was it still part of the RNA world? Although there are some reasons to consider the second hypothesis, the characteristics mentioned above suggest rather that LUCA's genome consisted of DNA. The differences between the three domains can be easily explained by DNA transfer from viruses that infected ancient cells.

Finally, LUCA multiplied by binary division, after duplicating all of its internal components.

Of course, LUCA was not only in its time, but coexisted with other organisms. Undoubtedly it shared a common ancestor with some of them (remember that LUCA is the last—the most recent—common ancestor of all living organisms, but other precursors, even older, undoubtedly existed; they are the ancestors of LUCA!) while others could have been derived from completely different evolutionary lines.[29]

The existence of LUCA allows us to elaborate the possible phylogenetic tree of living beings to evaluate their kinship ties. However, trees of this type require what is called, in jargon, the "out group". It is an "individual" outside the whole that is to be analysed: as far as possible, phylogenetically, from all the representatives who are part of the group. But if we want to study ALL living beings, we no longer have an out group! The construction of the tree becomes delicate…

Without going into technical details, it is necessary to specify that the result is subject to caution and that the various specialists present, in this regard, different points of view. The following presents the phylogenetic tree (Fig. 2.3) developed it in 1990 and proposed by Carl R. Woese, who discovered the archaea.

From LUCA starts a branch that leads to bacteria and a second that arrives at the common ancestor between archaea and eukaryotes. This means that we, humans, have a common ancestor, very distant phylogenetically, with the small *N. equitans*, which lives in Icelandic hot springs (do you remember it?

[29] According to some researchers, RNA viruses and certain DNA viruses could be the descendants of contemporary organisms of LUCA or even of beings that preceded him (Forterre 1996).

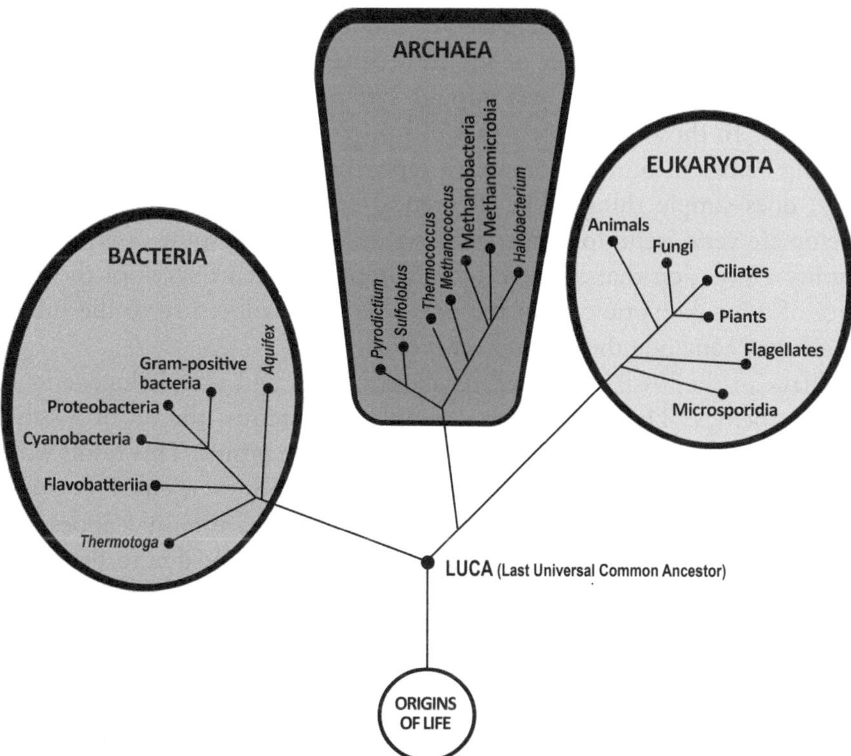

Fig. 2.3 The tree of living beings (bacteria, archaea and prokaryotes) proposed by Carl R. Woese, in 1990. In this representation, archaea and eukaryotes share a common ancestor that has already separated from bacteria. Redrawn from Forterre (2007)

The smallest living organism!). But to find an ancestor in common with *E. coli*, famous intestinal bacterium, we have to go back up the tree, in particular up to LUCA.

However, due to difficulties in establishing the "root" of the phylogenetic tree in the absence of out groups and the different hypotheses favoured by specialists studying the origin of Life, different possibilities have been proposed. One of these is to replace the branch of bacteria with that of eukaryotes. The latter, according to this hypothesis, would have differentiated themselves before the other two groups. In this case, archaea and bacteria would share a common ancestor at a time when nucleated cells had already appeared.

There is also another version, called the "ring of the living": LUCA would be the ancestor of bacteria and archaea. The eukaryotic cell would have arisen from the association between the representatives of the other two domains; at least one bacterium and one archaeon. Although I recognize the importance of the phylogenetic division of life into three domains, I still support the emergence of the eukaryotic cell through symbiotic association of prokaryotic organisms. To those who object that this proposal is ill-suited to the division of living beings into three domains, I reply that Nature never (or not necessarily) does simple things. The theoretical schematizations that researchers develop are very useful for understanding reality but, in some cases, they can simplify a situation that is much more complex. When I speak of the emergence of the eukaryotic cell, in the next chapter, I will return to the subject. Now, let us conclude the chapter on prokaryotes.

What can the fossil record tell us about LUCA and the phylogenetic tree of living beings? The answer is not simple. Apparently, the set of fossils of the Archean presents a rich array of prokaryotic organisms. This result would agree with various phylogenetic reconstructions. However, eukaryotes seem absent until the final part of the Archean. Therefore, it would appear that this group could not have emerged directly from LUCA, due to the lack of fossils. However, in palaeontology, the absence of evidence does not constitute evidence of absence. Perhaps we have not yet found what we are looking for, despite its presence. Perhaps they left no traces, or we have not been able to find them. It must also be considered that the Archaean sediments that can be explored are much less abundant than the more recent ones, contemporary with the dinosaurs or the Quaternary strata!

Before concluding this section, I wish to defend prokaryotes. These organisms (whether bacteria or archaea) are very old, but their descendants are still present among us. They are everywhere, around us (in the air, in the water, in the soil—tens or hundreds of meters underground) and even in our bodies: in the intestines, for example! We tend to think of these single-celled beings as very simple organisms, but their evolutionary history is longer than any animal, plant, or fungus.

They are also very numerous. Although it is difficult at present to put forward precise figures, bacteria would be the dominant group of all living beings, surpassing even insects, which are the most diverse group of animals (from one to several million!).

Bacteria and archaea are also two successful groups. Archaea are specialists capable of surviving in the most inhospitable environments[30]: temperatures close to those of boiling water, extremely high pressures, prohibitive salinity rates, anoxic environments.

[30] Of course, the term "survival" must be understood by human standards, because for archaea it is routine!

Bacteria constitutes an extremely diverse group (Stanier et al. 1963). Unfortunately, we tend to consider diversity in morphological and visual terms. So, we primarily focus on animals and plants. Indeed, we only notice the flashiest members of these groups! If their external form is not very indicative, bacteria constitute an even more diversified universe, in terms of behaviour and eating habits, than that of plants, fungi and animals combined. Imagine a molecule that includes carbon (excluding pure graphite): there is certainly a prokaryotic microorganism capable of feeding on it, using it as a source of carbon and metabolizing it to produce energy. Some bacteria can even derive energy from uranium reduction (Lovley et al. 1991), leading to proposals for their use in uranium bioremediation.

It is not surprising, therefore, that such numerous and different organisms have a greater impact on current ecological systems. This has undoubtedly been true in the past. An example? Among the oldest traces of living things are microfossils that, anatomically, remarkably resemble cyanobacteria, both in shape and in the type of associated deposits in which they are found. They are so similar that specialists attribute the same abilities to them. Among these, the most important is to carry out photosynthesis comparable to that of green plants. Since they appeared, these microorganisms have had to multiply and invade all available environments. In doing so, they expelled into their living environment the waste of their metabolism, gaseous oxygen, absent before. We will examine the effects of this environmental change later, one of the first "pollution event" in Earth biological history.

To those who claim that the world belongs to humans, zoologists respond that, in reality, the Earth belongs to insects. Biologists counter that this is incorrect: the world belongs to bacteria. And they are right!

2.4 What About Viruses?

Let us now consider another biological entity, which was already introduced at the end of paragraph 1.1. This is the "virus", whose status as a living organism is debated and even denied by some members of the international scientific community. I remind you that, according to the definition proposed by Mario Ageno, viruses are not living beings. At least under the traditional definition.

As discussed in paragraph 1.1, "viruses" are classically composed of genetic material, either DNA or RNA, in varying amounts (Crawford 2011). The genetic material is protected by a protein coat, the capsid, which prevents degradation of the genome as the "virus" travels from one host to another.

Some "viruses" also have an additional protective layer, an envelope that completely surrounded the capsid. The composition of this envelope is complex, including proteins, lipids and carbohydrates. It forms as the "viruses" leave the host to seek other targets and is composed of a mixture of elements of cellular origin, some from the host and others of viral origin.

The viral cycle begins with the infection of a host, continues with the production of viral copies, and ends with the release of the progeny, destroying the parasitized cell. One released, the cycle repeats in the same manner. Upon reaching its target, the "virus" penetrates the host cell and may hijack its cellular machinery to produce viral components from viral genome. Some "viruses", such as herpes simplex virus (HSV), can remain quiescent for a long time, sometimes until the host's death. Under certain conditions, as weakened immunity, they reactive. The viral genome can also integrate into the host genome, adding new DNA sequences to the host chromosomes (Watson et al. 1987).

But their skills don't stop there. Some "viruses" can even transport fragments of the host's DNA to new cells, enabling horizontal gene transfer between individuals of the same generation. Clearly, "viruses" have multiple strategies at their disposal!

"Viruses" infect all three domains of Life: eukaryotes, bacteria and even archaea, including those inhabiting the most extreme environments; "viruses" haunt them and infect them even down there!

"Viruses" are extraordinarily abundant, possibly the most numerous biological entities on Earth. In the upper layers of the oceans, viral abundance may exceed bacteria counts tenfold (Forterre 2007, p. 138, 2010). This is partly because each cell can be infected by multiple "virus".

A full exploration of viral diversity and properties is beyond the scope of this section, but notable examples include mimiviruses (whose name derives from 'mimicking microbe virus' [= microbe imitating a virus] because they were initially mistaken for prokaryotes). Mimiviruses (Fig. 2.4) have icosahedral capsids about 400 nm in diameter, comparable to small prokaryotes and genome encoding 600–1000 genes. They infect amoebae, such as *Entamoeba haemolytica.*

Another example is the SSV1 "virus" (*Sulfolobus shibatae* virus n°1; Fig. 2.4), which infests the hyperthermophilic archaeon of the same name, living in a Japanese hot spring. Despite extreme conditions, these "viruses" successfully infect their host. SSV1 "virus" was the first to be isolated in an archaea population (Forterre 2007, pp. 207–211). But now, these discoveries are multiplying.

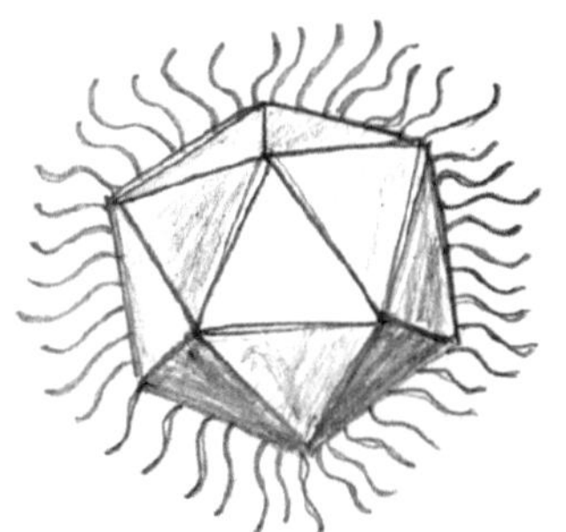
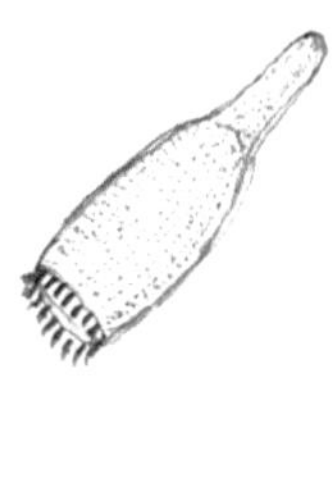

Fig. 2.4 On the left, the virion of a bacteriophage (dimensions between 25 and 200 nm); in the centre, the virion of mimivirus, whose icosahedral capsid measures, without fibrils, about 400 nm (this virion infects the amoeba *Entamoeba haemolytica*); on the right, the virion of *Sulfolobus shibatae* virus n°1, with a bottle-shaped capsid, which attacks the hyperthermophilic archaeon of the same name (size, 150 nm approximately). The figured virions are not represented to scale

Now let us consider the nature and evolutionary significance of "viruses". Why are they important in the history of Live? Classical definition considers "viruses" non-living because they lack cellular machinery for independent biological activity. However, recently, the conception of these biological entities has changed. The definition of the virus has also changed. Some researchers now consider them to be genuine living beings. But why? And Ageno's definition, then?

The entities classically called viruses do not correspond to living beings for the already mentioned reasons. However, this entity does not constitute the actual virus, but just the virion, the infectious agent, which propagates from one cell to another. A kind of "spore" or "spermatozoon" that ensures the propagation of the species. The true living entity is another. That's why I have always written the name in quotation marks.

When a cell is infected, the viral genome takes control of the host and a new biological entity is formed: it contains the viral genetic heritage plus the host's DNA. A part of the latter is degraded, because the invader is only interested in expressing its viral genes.

This new entity would be the real virus. The "viral cell": a host cell hijacked by an active viral genome. In this state the viral cell satisfies the criteria of a living organism, as it executes cellular functions under the control of the viral genome, rather than its original host program.

Professor Forterre (2010, 2011) proposed that viral cell (host cell + active viral genome) be called a virus, while the virion (which is the entity corresponding to the classic definition of "virus") be just considered like a propagation agent. This perspective emphasizes that viruses are extreme parasites that have evolved a unique mode of reproduction. Instead of cell division or sexuality,[31] viruses use a vector, the virion, to transfer their genomes from host to host, manifesting their organismal functions within the hijacked cell. Of course, all this is implemented with multiple variations on the central theme, but the basic concept remains the same.

Would you like additional proof? The giant virus mamavirus (a "relative" of mimiviruses) produces a spherical complex (a "virus factory") inside its victims[32] to assemble new virions. This factory can even be infected by another virus, dubbed Spoutnik (Forterre 2011). A viral cell attacked and infected by another virus! Yet only living beings are susceptible to viral attack!

Despite this revised definition, virions remain essential for characterizing and recognizing viruses.

The origin of viruses is still debated. Some hypotheses suggest they descend from ancient cellular organisms that developed extreme parasitism. Early victims were other unicellular organisms. Viruses may have evolved to inject their genome into host cells, producing new copies while minimizing the need to replicate structural components independently. This led to the contemporary viral cycle, alternating between the viral cell (the entity responding to the definition of the living being) and the virion.

But what is the interest of the virus in the history of biological evolution? Viruses would be very ancient; they may predate LUCA (Forterre 2006; Forterre et al. 2005) and could have existed during the hypothesized RNA world, being the only remaining witnesses. Not only that, but they would have favoured the transition from the RNA to the DNA world.[33]

Some researchers even propose that viruses played a role in the formation of the eukaryotic nucleus (Forterre 2011). Here is an interesting and original hypothesis, capable of positively integrating the theory of eukaryotic cell formation, formulated by Lynn Margulis, which we will deal with in the next chapter.

---

[31] I will present sexuality as the opposite of cell division in Chap. 3, dedicated to eukaryotes.

[32] The constitution of these complexes, the "Virus factories", where the different elements are synthesized to assemble the virions, would be a characteristic of viral infections on eukaryotes. These complexes allow the virus to be visualized as a living entity different from the host cell. In the case of mimivirus infection, these complexes are so large that they can be seen under an optical microscope. In bacteria and, in general, in all prokaryotes, such structures do not form. In this case, the whole host cell is the "Virus factory".

[33] If we accept that viruses, as defined by Patrick Forterre, are real living beings and if we believe they derive from organisms existing at the time of the hypothetical RNA world, then we should review our definition of the common ancestor. LUCA, in this case would be the last common ancestor to the ONLY members of the three domains established by Woese, but not to viruses.

This hypothesis is attractive, though currently speculative, because virions do not fossilize (or do so very rarely). However, these speculations, advanced to explain observations and experimental data, undoubtedly constitute excellent starting points to feed future research on issues related to the early stages of the origin of Life on Earth.

Although speculative due to the rarity of viral fossil, the importance of important evolutive virus roles is supported by observations and experimental data. Approximately 13% of bacterial and archaeal genomes consists of sequences of viral origin, transported and integrated into the DNA of prokaryotes by virions. This proportion would even rise to 40% in humans (Forterre 2011; Forterre and Gaïa 2018). Could we be naturally occurring GMOs shaped by viral evolution?

2.5 Summary of This Chapter

The recognition of traces of ancient Life in Archean sediments is the result of meticulous and rigorous chemical analyses (particularly of isotope ratios, especially carbon) and/or anatomical and morphological comparisons with well-defined samples. Usually, both types of analyses are required to obtain reliable results.

The oldest confirmed traces, accepted by most of the scientific community, date back 3.5 billion years and are found in sedimentary layers in African and Australian. These include stromatolitic structures of varying size and isolated microfossils. These traces testify to the presence of diverse microscopic creatures with different metabolic strategies. Among these were photosynthetic organisms, close to modern cyanobacteria, as well as other microbes, some living near hydrothermal vents or capable of synthesizing methane, such as methanogenic archaea. All of them lived in water, which provided protection from harmful cosmic radiation that was not yet attenuated by an ozone layer (O_3), which was completely absent at the time.

The presence of photosynthetic organisms supports Mario Ageno's hypothesis regarding the emergence of Life.

Current data indicate that, during the early Archean, biodiversity included various types of prokaryotic microorganisms: bacteria and archaea. Eukaryotes appeared later, about 2.7 billion years ago. By the end of the Archean, the three domains of Life were already present on Earth. Fossil or biological evidence consistently indicates that all modern organisms, as well as those that lived in the past, share a common ancestor called LUCA. Even if Life arose more than once, at different times or locations, only LUCA's lineage persisted. Among his descendants are all modern organisms, as well as those we will discuss in these pages.

Viruses may represent remnants of organisms that existed before LUCA or that were on contemporaneous with it. These entities could be ancient organisms that pushed parasitism to the extreme, gaining the ability to hijack the metabolic machinery of infected cells and force them to produce copies of themselves. Over evolution time, rather than creating complete copies, viruses limited themselves to producing their genome enclosed within a protective protein coat. The vectors, the virions, served to preserve the viral genome during dispersion and infection of new cells.

Thus, viruses may be considered unique living beings: the viral cell (the host infected by the virus genome) represents the actual living organism, while the virion is merely a vector for transferring genetic material between cells. Despite this distinction, viral identification and classification still rely on the virion.

The relationship between viruses and other organisms may be more intimate than the simple host-parasite interaction suggests. Viruses may have facilitated the transition from the hypothetical RNA world to the DNA world, thereby driving a decisive evolutionary change. RNA, as the repository of genetic information, would have persisted only in some viruses. Even the nucleus of eukaryotic cells may have arisen due to viral influence.

However, because virions rarely fossilize, much of this remains speculative. Nevertheless, these hypotheses provide excellent foundations for future research into the early evolution of Life.

Appendix

Mimivirus, the Giant "Virus"

Some giant "viruses" possess genomes with more than 500 genes. Mimiviruses (Fig. 2.4) have between 600 and 1000 genes, exceeding that of some bacteria. Their DNA resembles that of eukaryotes more than prokaryotes. Remarkably, mimiviruses carry genes encoding proteins involved in DNA repair and the translation of RNA into proteins. This is unusual because these functions already exist in the host's cellular machinery, which is fully exploited to for the production of mimivirus virions (Crawford 2011).

References[34]

Ageno, M. 1991. *Dal non vivente al vivente*. Rome-Naples: Theoria.

Allwood, A.C., M.R. Walter, B.S. Kamber, C.P. Marshall, and I.W. Burch. 2006. Stromatolite reef from the Early Archaean era of Australia. *Nature* 441: 714–718.

Allwood, A.C., M.R. Walter, I.W. Burch, and B.S. Kamber. 2007. 3.43 billion-year-old stromatolite reef from the Pilbara Craton of Western Australia: Ecosystem-scale insights to early life on Earth. *Precambrian Research* 158: 198–227.

Awramik, S.M. 2006. Respect for stromatolites. *Nature* 441: 700–701.

Bertrand-Sarfati, J., P. Freytet, and J.-C. Plaziat. 1966. Les calcaires concrétionnés da la limite Oligocène-Miocène des environs de Saint-Pourçain-sur-Sioule (Limagne d'Allier): rôle des algues dans leur édification; analogue avec les stromatolites et rapports avec la sedimentation. *Bulletin de la Société géologique de France* 8: 652–662.

Brasier, M.D., O.R. Green, A.P. Jephcoat, A.K. Kleppe, M.J. Van Kranendonk, J.F. Lindsay, A. Steele, and N.V. Grassineau. 2002. Questioning the evidence of Earth's oldest fossils. *Nature* 416: 76–81.

Calvin, M. 1969. *Chemical evolution*. Oxford: Clarendon Press.

Crawford, D.H. 2011. *Viruses—A very short introduction*. Oxford University Press.

Dodd, M.S., D. Papineaua, T. Grennec, J.F. Slackd, M. Rittnerb, F. Pirajnoe, J. O'Neilf, and C.T.S. Little. 2017. Evidence for early life in Earth's oldest hydrothermal vent precipitates. *Nature* 543: 60–64.

Fedo, C.M., and M.J. Whitehouse. 2002. Metasomatic origin of quartz-pyroxene rock, Akilia, Greenland, and implications for Earth's earliest life. *Science* 296: 1448–1452.

Forterre, P. 1996. À la Recherche de LUCA. In *Meeting Fondation de Treilles*.

Forterre, P. 2006. The origin of viruses and their possible roles in major evolution transitions. *Virus Research* 117: 5–16.

*Forterre, P. 2007. *Microbes de l'enfer*. Paris: Belin pour la Science.

Forterre, P. 2010. Defining Life: The virus viewpoint. *Origin of Life and Evolution of the Biosphere* 40 (2): 151–160.

Forterre, P. 2011. Manipulation of cellular syntheses and the nature of the viruses: The virocell concept. *Comptes Rendus Chimie* 14 (4): 392–399.

Forterre, P., and M. Gaïa. 2018. La place des virus dans le monde vivant: le concept de virocell. In *Microbiodiversité. Un nouveau regard*, ed. L. Palka, 23–49. Paris: Matériologique.

Forterre, P., S. Gribaldo, and C. Brochier. 2005. Luca: À la recherché du plus proche ancêtre commun universel. *Médicine Sciences* 21 (10): 860–865.

[34] Works preceded by an asterisk can be read, more or less easily, by people with a non-professional skilling.

Freytet, P. 2000. Distribution and palaeoecology of non marine algae and stromatolites: II, the Limagne of Allier Oligo-Miocene lake (central France). *Annales de Paléontologie (Vertébrés et Invertébrés)* 86 (1): 3–57.

Freytet, P., and E.P. Verrecchia. 1998. Freshwater organisms that build stromatolites: A synopsis of biocrystallization by prokaryotic and eukaryotic algae. *Sedimentology* 45 (3): 535–563.

*Fry, I. 2000. *The emergency of Life on Earth—A historical and scientific overview.* New Brunswick (New Jersey): Rutgers University Press.

Hoffman, P. 1967. Algal stromatolites: Use in stratigraphic correlation and paleocurrent determination. *Science* 157: 1043–1045.

Holland, H.D. 1997. Evidence for Life on Earth more than 3850 million years ago. *Science* 275: 38–39.

Javaux, E.J., C.P. Marshall, and A. Bekker. 2010. Organic-walled microfossils in 3.2-billion-year-old shallow-marine siliciclastic deposits. *Nature* 463: 934–939.

*Knoll, H.A. 2004. *Life in a young planet—The first three billion years of evolution on the Earth.* Princeton University Press.

Koonin, E.V. 2003. Comparative genomics, minimal gene-sets and the last universal common ancestor. *Nature Reviews Microbiology* 1: 127–136.

Lindsay, M.R., R.I. Webb, M. Strous, M.S.M. Jetten, M.K. Butler, R.J. Forde, and J.A. Fuerst. 2001. Cell compartmentalisation in planctomycetes: Novel types of structural organisation for the bacterial cell. *Archives of Microbiology* 175: 413–429.

Lovley, D.R., E.J.P. Phillips, Y.A. Borby, and E.R. Landa. 1991. Microbial reduction of uranium. *Nature* 350: 413–416.

Mojzsis, S.J., G. Arrhenius, K.D. McKeegan, T.M. Harrison, A.P. Nutman, and C.R.L. Friend. 1997. Evidence for life on Earth before 3800 million years ago. *Nature* 384: 55–59.

Muir, M.D., R.B. Grant, G.M. Bliss, W.L. Dives, and D.O. Hall. 1977. A discussion of biogenicity criteria in a geological context with examples from a very old greenstone belt, a Late Precambrian deformed zone, and tectonized Phanoerozoic rocks. In *Chemical evolution of the Early Precambrian*, ed. C. Ponnamperuma, 105–170. New York: Academic Press.

Nisbet, E. 2000. The realms of Archaean life. *Nature* 405: 625–626.

Nutman, A.P., V.C. Bennett, C.R.L. Friend, M.J. Van Kranendonk, and A.R. Chivas. 2016. Rapid emergence of life shown by discovery of 3,700-million-year-old microbial structures. *Nature* 573: 535–538.

Penny, D., and P. Poole. 1999. The nature of the last universal common ancestor. *Current Opinion in Genetics and Development* 9 (6): 672–677.

Rasmussen, B. 2000. Filamentous microfossils in a 3,235 million-year-old volcanogenic massive sulphide deposit. *Nature* 405: 676–679.

Rosing, M.T., and R. Frei. 2004. U-rich Archaean sea-floor sediments from Greenland—Indications of > 3700 Ma oxygenic photosynthesis. *Earth and Planetary Science Letters* 217: 237–244.

Schopf, J.M., and B.M. Parcker. 1987. Early Archean (3.3 billion to 3.5 billion-year old) microfossil from Warrawoona Group, Australia. *Science* 237: 70–73.

Schopf, J.W., A.B. Kudryavtsev, A.D. Czaja, and A.B. Tripathi. 2007. Evidence of Archean life: Stromatolites and microfossils. *Precambrian Research* 158: 141–155.

Shen, Y., R. Buick, and D.E. Canfield. 2001. Isotopic evidence for microbial sulphate reduction in the early Archaean era. *Nature* 410: 77–81.

Stanier, R.Y., M. Doudorof, and E.A. Adelberg. 1963. *The microbial world*, 2nd ed. Prentice-Hall.

Taylor, T.N., E.L. Taylor, and M. Krings. 2009. *Paleobotany: The biology and evolution of fossil plants*, 2nd ed. Amsterdam: Academic Press.

Tice, M.M., and D.R. Lowe. 2004. Photosynthetic microbial mats in the 3,416 Myr-old-ocean. *Nature* 431: 549–552.

Ueno, Y., K. Yamada, N. Yoshida, S. Maruyama, and Y. Isozaki. 2006. Evidence from fluid inclusions for microbial methanogenesis in the Early Archaean era. *Nature* 440: 516–519.

Van Zuilen, M.A., A. Lepland, and G. Arrhenius. 2002. Reassessing the evidence for the earliest traces of life. *Nature* 418: 627–630.

Wacey, D., M.R. Kilburn, M. Saunders, J. Cliff, and M.D. Brasier. 2011. Microfossils of sulphur-metabolizing cells in 3.4-billion-year-old rocks of Western Australia. *Nature Geoscience* 4: 698–702.

Watson, J.D., N.H. Hopkins, J.W. Roberts, J. Argetsinger Steitz, and A.M. Weiner. 1987. *Molecular biology of the gene*, 2 vol., 4th ed. San Francisco: The Benjamin/ Cummings Publishing Company.

Wattinne, A., E. Vennin, and P. De Wever. 2003. Evolution d'un environnement carbonaté lacustre à stromatolithes, par l'approche paléo-écologique (carrière de Montaigu-le Blin, bassin des Limagnes, Allier, France). *Bulletin de la Société Géologique de France* 174 (3): 243–260.

Williams, T.A., P.G. Foster, C.J. Cox, and T.M. Embley. 2013. An archaeal origin of eukaryotes supports only two primary domains of life. *Nature* 504: 231–236.

Woese, C.R., and G.E. Fox. 1977. Phylogenetic structure of the prokaryotic domain: The primary kingdoms. *Proceedings of the National Academy of Sciences U.S.A.* 74 (11): 5088–5090.

Woese, C.R., O. Kandler, and M.L. Wheelis. 1990. Towards a natural system of organisms: Proposal for the domains Archaea, Bacteria, and Eucarya. *Proceedings of the National Academy of Sciences U.S.A.* 87 (12): 4576–4579.

3

Eukaryotes Enter the Scene

3.1 A New Kind of Cell

In the previous chapter, we discussed the oldest known living organisms. Based on their remains, I attempted to reconstruct a picture of the Archaean ecosystems of our planet. I also emphasized that prokaryotic biodiversity was significant in the past, and still is today. These microorganisms continue to play a predominant role in the biosphere. Throughout our history, I will return to this concept several times.

However, since the end of the Archean, prokaryotes have not been alone; other organisms enter the stage. It is time to introduce them and uncover the surprises that the history of Life holds once they appear.

Beyond the 2.5 Ga boundary, we are in the Proterozoic, the time interval between this date and about 541 million years (Ma). This is the period we are now concerned with. The world was still dominated by prokaryotes, who continue to evolve and diversify. Yet it is also the time when the third domain of Life made its appearance. Although it first emerged in the final stage of Archean, it became distinct only in the following eon, the Proterozoic.

Let us begin by examining the characteristics of these new protagonists, the eukaryotes. Among their distinctive features, we can mention (see also Fig. 3.1):

A. M. F. Valli, *The Three Domains of Life*, Copernicus Books,
https://doi.org/10.1007/978-3-032-14802-5_3

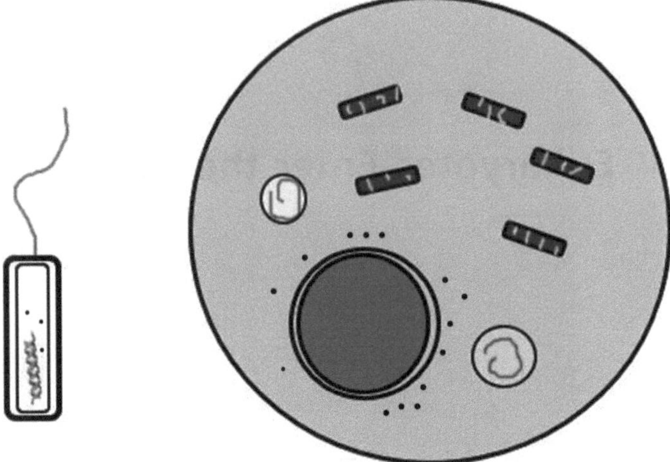

Fig. 3.1 Main differences between the prokaryotic cell (the smallest one on the left) and the eukaryotic cell (the largest, spherical on the right). In prokaryotes, DNA is located directly in the cellular cytoplasm (the internal liquid environment); in eukaryotes, DNA is enclosed within the nucleus (the darker spherical structure inside the eukaryotic cell). The eukaryotic cell also contains specialized organelles: mitochondria (dark, rod-like shapes with striped patterns) and plastids (lighter spheres, present only in plant cells). The black dots, present in both cell types, represent the ribosomes. The two cells are not shown at the same scale

- the presence of a real nucleus[1] (a double membrane consisting mostly of lipids, such as those already described in Chap. 1) enclosing the genetic material;
- the possession of specialized organelles (such as mitochondria, found in all cells, and chloroplasts, in present in photosynthetic organisms[2]);
- structural and organizational differences at the genetic level within DNA.[3]

[1] Athough ALL eukaryotes possess a double-membrane nucleus containing cellular DNA, this may not be an exclusive feature of the group. Indeed, certain prokaryotes of the phylum Planctomycetota have similar structures surrounding their genome. See footnote 16 in Chap. 2.

[2] A notable feature of mitochondria and plastids (the group of organelles to which chloroplasts belong) is that they contain DNA in them. This DNA is essential for organelle (which occurs by binary fission, like bacteria) but it is not sufficient to complete the process. This means that mitochondria and chloroplasts do not possess the complete genetic information to produce a full copy of themselves. They have only a part of it; the rest resides in the nucleus of the cell. Conversely, the nuclear genome alone cannot ensure organelle duplication; the necessary information is split between the nucleus and the organelles.

[3] One of the most important features of the eukaryotic genome is the presence of introns within protein-coding DNA sequences. Introns are transcribed into messenger RNA (mRNA), but are removed before translation at the ribosomes (a process known as mRNA "splicing"; Watson et al. 1987).

In my view, the eukaryotic cell represents a genuine evolutionary "novelty" compared to prokaryotes. In fact, it is capable of achievements far beyond the reach of bacteria and archaea.[4] For instance, eukaryotic cells are usually larger than their "cousins." The latter generally measure only a few micrometres (μm) in length or diameter,[5] although some spirochetes (helically shaped microbes responsible for numerous diseases in humans and animals) can reach up to 500 μm. By contrast, nucleated cells are typically ten times larger, averaging between 10 and 100 μm. While there is some overlap, eukaryotic cells grow too much greater sizes than those of prokaryotes.

The largest eukaryotic cells are found in the animal kingdom. The ostrich (*Struthio camelus*) is, in terms of mass, the largest living cell, weighing more than a kilogram. If consider length instead of mass, the record goes to the giant squid neuron (*Architeuthis dux*), whose axon, one of its extensions, exceeds 10 m! Clearly, the largest eukaryotic cells far surpass even the biggest prokaryotic ones. It is almost like comparing objects from entirely different worlds, much like comparing Luxembourg to the United States of America. Both are both sovereign countries, but the vast size of the latter makes the former appear negligible!

The differences are not limited to size. Eukaryotes can perform functions unknown to bacteria and archaea. They are capable of forming multicellular organisms, composed of many different cells working together in a coordinate way. By contrast, A collection of prokaryotic cells remains a loose association, each cell maintaining its individuality. The eukaryotic cell is therefore fundamentally different from its prokaryotic counterpart.

That said, we should not underestimate bacteria or archaea. These organisms are capable of feats no less remarkable. True, they are small and unable to form genuine multicellular beings. But prokaryotes have managed to colonize virtually every environment on Earth: the deep surface, the ocean depths, hydrothermal vents, and even polar snowfields. They have also invaded the body of the most complex organisms. By comparison, the habitats of eukaryotes, whether unicellular or multicellular, represent only a fraction of those occupied by prokaryotes. Often, bacteria or archaea colonization enables other organisms, such as plants or animals, to settle in turn. For instance, at hydrothermal vents, the metabolic abilities of microorganisms make it possible for animals living in symbiosis with them to survive in otherwise hostile environments.

Whereas prokaryotes have occupied most natural environments while remaining small, eukaryotes acquired very different properties, allowing them

---

[4] If you have forgotten what an archaeon is, please refer back to Sect. 2.2.

[5] A micrometre (μm) is one millionth of a metre. See paragraph 2.2.

to exploit other ecological niches. Modern biodiversity would be vastly diminished without the coexistence of both cell types!

Now let us compare the functioning of the eukaryotic cell with that of the prokaryotic one. An analogy may help clarify the difference. Imagine a small travel agency in an important city neighbourhood. The manager (man or woman) is the sole employee, handling everything: answering calls, welcoming customers, and organizing trips. The clientele is modest, but the business prospers enough for the manager to make a decent living.

After some time, the manager hires an assistant with specific tasks: answering phone calls, noting customer requests, and passing the information one. Free from routine calls, the manager can now develop new activities: building relationships with other agencies, exploring new destinations, or reaching new clients. Thanks to the assistant, the agency's activity expands, even though there is an additional cost: the new salary. Over time, the agency may grow further, hire more staff, and eventually open new branches in other neighbourhoods or cities.

This analogy illustrates the differences between prokaryotic and eukaryotic cell. The prokaryotic cell is like the one-person agency, where a single entity performs all tasks The eukaryotic cell, by contrast, distributes functions among specialized components, making new activities possible (such collaboration or expansion).

The prokaryotic strategy (though simplified here) can be summarized as follows: the cell adapts to a particular environment, devotes all its energy to growing and dividing, and leaves as many descendants as possible. When conditions change (e.g., the appearance of pollutants or depletion of nutrients), the microorganism halts growth and waits. In time, it either perishes or, if some descendants are beginning better adapted, colonized the new environment. The vast diversity of environments occupied by bacteria and archaea, and their immense variety today, testify to their extraordinary resilience.

Eukaryotes, however, are not competitive on this ground. They therefore developed new strategy: division on labour, especially in energy production. This task is delegated to mitochondria (and a bit to chloroplasts, in plants). While organelles are costly to maintain, they perform functions more efficiently than the centralized system of prokaryotes. The eukaryotic system thus produces a small energy surplus, with can be diverted away from basic growth and division. That surplus can fuel innovation, for example, an increase in cell size! Larger cells can exploit resources inaccessible to smaller prokaryotes. Some can even engulf and digest them!

Surplus energy can also be invested in creating specialized molecules for "communication" between cells. This is the kay to further evolutionary advances, as will be explained in the next section. For now, let us turn to another problem: how did the eukaryotic cell arise from the prokaryotic world?

The most widely accepted theory is that the eukaryotic cell originated from a symbiotic association[6] between prokaryotic organisms. According to this idea, microbes from different groups, complementary in function, joined forces and began to share essential tasks. Together, they gave rise a new organism, capable of achievements beyond those of any individual prokaryote. The hypothesis was pioneered by Lynn Margulis, an American biologist at Berkeley. Her original model proposed a three-stage symbiosis involving four different prokaryotic organisms.[7]

(1) A thermophilic, sulphur-reducing archaeon (do you remember that some chemical traces in the archaean layers testified to the existence of sulphur-reducing microorganisms starting from this eon?) associated with a mobile, spirochete-like. This union formed the body of a new entity, with a nucleus and a mobile appendage (the flagellum), but still lacking organelles.
(2) Later, mitochondria arose from incorporation of an oxygen-consuming bacterium, possibly related to rickettsiae (endoparasitic microorganisms).
(3) Finally, chloroplasts would derive from cyanobacteria.[8]

The symbiosis between the first three organisms would have allowed the formation of a new heterotrophic cell. The addition of one or more cyanobacteria would have allowed him an additional possibility: the possibility of carrying out photosynthesis in the same way as the assimilated bacterium (Fig. 3.2). I would like to point out, once again, that green plants and cyanobacteria carry out a similar photosynthesis, which produces oxygen as a

[6] Symbiosis is a stable association between partners of different species, providing benefits to all.

[7] Margulis (1999); de Reviers (2018). For a more recent synthesis, consult Archibald's work (2014). His book emphasizes the importance of endosymbiosis for the constitution of eukaryotic cells and details the evolutionary stages. A relatively recent model to explain endosymbiosis leading to eukaryotes was proposed by Baum and Baum (2014). See also its development in Mieli et al. (2025).

[8] Lynn Margulis' theory did not consider the intervention of viruses for the formation of the nucleus. At that time, "viruses" still corresponded to the classical definition introduced towards the end of paragraph 1.1. They were not yet considered living beings. The ideas that begin to involve them in the evolution of prokaryotes and eukaryotes will only develop later. For now, I will limit myself to describing the theory of Margulis and the concepts that have been accepted, without adding the details and the latest innovations concerning the evolution of eukaryotes. To understand the theory of Margulis and the origin of mitochondria and chloroplasts, I invite you to read Chap. 9 of the book by Selosse (2017). However, the subject is in full "effervescence". For example, in a relatively recent article (Degli Esposti et al. 2014), an alternative method of analysis is proposed to discover the possible ancestors of mitochondria, resulting in a different solution than that reported in the text.

waste element. Lynn Margulis' theory allows us to understand why the two photosyntheses are analogous!

How plausible is this scenario? What evidence supports it? Most researchers reject the incorporation of a spirochete. So, the prevailing view is that eukaryotic cells originated from three microbes:

- an archaeon microorganism (providing the main body);
- protobacteria (as rickettsiae, ancestors of mitochondria);
- cyanobacteria, (ancestors of chloroplasts).

Biochemical analysis and genome sequencing support this. Organellar genomes are closely related to protobacteria (mitochondria) and cyanobacteria (chloroplasts). The fact that organelles contain their own DNA, distinct from nuclear DNA, and cannot be regenerated if lost, is strong evidence of their exogenous (external) origin.

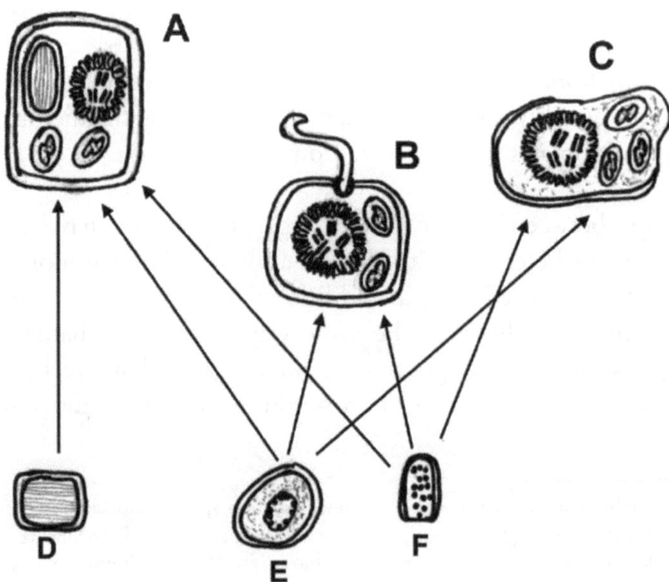

Fig. 3.2 Schematic representation of Margulis' theory. Animal cells (B) derive from a symbiotic association between a thermophilic prokaryote (E), providing the cell body, and aerobic bacteria (F), which become mitochondria (small ovals with inner dots). Fungal cells (C) share the same origin. Plant cells (A), added cyanobacteria (D), which become chloroplasts (rectangular figures with lighter bars). The nuclei are rounded with wavy edges and contain chromosomes. Margulis also proposed that spirochete (or similar microorganism) contribute to forming the flagellum

Similarities between eukaryotic molecules and archaeal homologues also confirm archaeal involvement. Thus, the eukaryotic cell appears to be a true product of prokaryotic symbiosis, remarkably one involving both domains of Live (Bacteria and Archaea: do you remember the "ring of the living", in Sect. 2.3?). The nucleated cell therefore represents a further evolutionary step beyond the enucleated cell.[9] It also provides one of the first examples of stable symbiosis as driving force in evolution. I remind the importance, emphasized in the introduction, of this type of interaction to appreciate biodiversity.

Nevertheless, Margulis' theory, even in its streamlined form (three organisms instead of four), is not universally accepted. The recognition of three domains (Bacteria, Archaea, and Eukaryotes) raised new questions. How did eukaryotic-specific proteins and structures arise from the simpler prokaryotic systems? Some researchers argue they evolved within a unique lineage of eukaryotes, without direct input from the other two domains (Forterre et al. 2005).

Others suggest LUCA itself may have had intermediate cell type, or that the nucleus originated from viral involvement (Forterre 1996; Forterre et al. 2005; Forterre and Gaïa 2018; do you remember the ideas from Sect. 2.4 about the role of viruses on the formation of the cell nucleus? So, why not consider a symbiosis between prokaryotes and viruses?).

As research progresses, new discoveries will likely refine or even overturn current models. A synthesis of Margulis' theory with alternative hypotheses is also possible (for example, combining prokaryotic symbiosis with a viral contribution to nucleus formation). Only through deeper study of all three domains, and a precise understanding of their similarities and differences, will a comprehensive theory of cell origins emerge.

[9] My mentor, Mario Ageno, when I was student at the University of Rome, used to tell us that the greatest evolutionary divide lies between cells without a nucleus and those with one. Of course, writing this text, I remain influenced by his words.

3.2 Multicellular Organisms and the Invention of Sexuality

Weighing more than a kilogram, the ostrich egg[10] is undoubtedly an imposing cell, whose size exceeds all others, whether prokaryotic or eukaryotic. However, even larger cells have existed in the past.[11] For example, the egg of the Madagascar elephant bird (*Aepyornis maximus*) measured one meter in circumference and had a volume equal to that of 160 chicken eggs! This bird would disappear recently (even after the year 1 000), seemingly due to human colonization of the island. Its fossil remains are well known: not only bones but also egg remains, some of which are complete.

However, if we look around, the life forms we see are huge, much larger than any giant bird or even dinosaur egg! Animals and plants, which form the biodiversity visible to our eyes, vary in size, ranging from a few millimetres and a few milligrams up to a height of 150 m for the giant sequoia (*Sequoiadendron giganteum*), the living organism that possesses, in its individual state, the most impressive size, with a corresponding mass.

The growth of eukaryotes is not limited to simply "enlarging" cell size. These organisms have acquired a new ability apparently inaccessible to prokaryotes: they are able to "produce" complex individuals, each made up of many different cells. With the emergence of eukaryotes, multicellular beings make their entrance into the history of Life!

But how is it possible that a solitary single-celled organism can associate with others of the same species to form a more complex organism? If the association between different organisms can be favoured by their complementarities, that between individuals of the same species is more difficult to understand, because they would rather compete for space and resources. Rather than coming together, they would distance themselves in a way that minimizes competition. Not only that, since similar beings produce the same waste, the concentration of these substances would tend to make conspecifics move away rather than come together. Nothing, therefore, seems to favour the association of closely related organisms.

However, various examples, all deriving from current reality, show us how this stage could have been overcome in the past. In this regard, I invite

---

[10] It is necessary to specify that the real cell is the egg yolk; the rest, including the shell, are only protective structures and packaging.

[11] Carnivorous dinosaurs, such as those of the family of Tyrannosauridae (genera *Tyrannosaurus* and *Tarbosaurus*) laid huge eggs, among the largest known. These eggs, weighing several kilos, had an elongated shape measuring up to 50 centimetres. In contrast, sauropods, those giants with oversized neck and tail (for example, *Diplodocus* or *Brachiosaurus*), laid proportionally smaller spherical eggs.

you to study the behaviour of a modern amoeba known as *Dictyostelium discoideum.*[12]

As noted in the previous paragraph, the eukaryotic cell, thanks to the division of tasks and, in particular, to the concentration of energy production in the mitochondria, has more significant energy flows than those of prokaryotes. It can, therefore, afford to "invest" in a certain number of innovations. Among these, the production of special molecules which, scattered outside the cell, act as real signals for all related microorganisms. *Dictyostelium discoideum* has learned to use these signals in a very precisely. Let us examine its behaviour in more detail (Fig. 3.3).

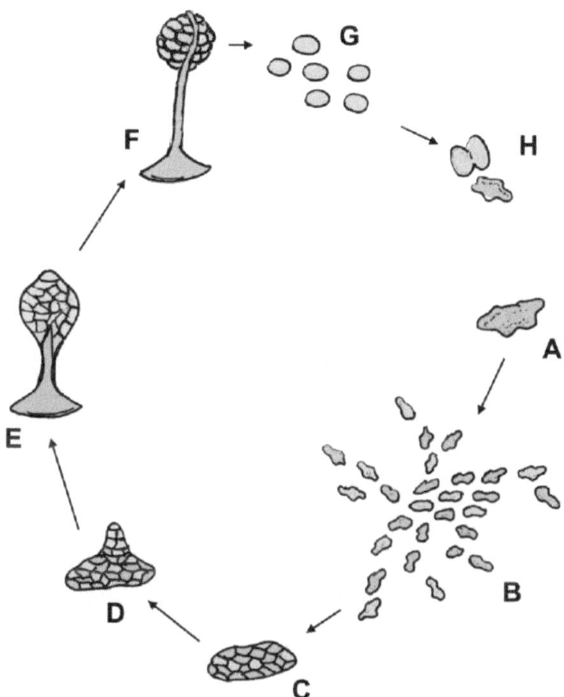

Fig. 3.3 The life cycle of *Dictyostelium discoideum* under conditions of dietary stress: (A) amoeba under normal conditions, (B) amoebas aggregate following the emission of cAMPs by one of them; (C) formation of pseudo-plasmodium; (D) cellulose secretion by cells on top of the pseudo-plasmodium; (E) formation of the stem and the fruiting body at its top; (F) maturation of the fruiting body; (G) spore release; (H) new generation of amoebas emerging from the spores. The cycle then can repeat

[12] *Dictyostelium discoideum* belongs to a group of slime moulds (myxomycetes). These eukaryotes are amoebas characterized by a particular life cycle, which manifests itself under nutritional deficiencies. This cycle is explained in the text and illustrated in Fig. 3.3.

This amoeba[13] is a very common organism on the forest floor of North America. Normally, it leads a solitary existence, seeking its nourishment on the ground (mainly bacteria, but also fungi). However, when food becomes scarce, the organism changes its behaviour, releasing a particular chemical compound, cyclic adenosine monophosphate (cAMP)[14] which acts as a "lure agent". All *Dictyostelium* that come into contact with cAMP begin to move towards the source of the emission, releasing a certain amount of this chemical compound themselves. In this way, the effect is amplified. As the signal strengthens, all amoebas in the surrounding area begin moving toward the first one that triggered the process.

Once all the *Dictyostelium* cells have reunited, they assemble into a gelatinous mass called a pseudo-plasmodium, which begins to move as if it were a single individual. Initially, the pseudo-plasmodium wanders randomly. Then, at a certain point, the cells located on top start secreting cellulose. Several cells then migrate toward the highest point, forming, as they rise, a stem on the back of the pseudo-plasmodium. The stem, reinforced by secreted cellulose, grows taller, and a fruiting body forms at its top.

Once the pseudo-plasmodium is formed, the different amoebas cease individual behaviour. From that moment, all actions are coordinated, and each cell specializes in a task (stem formation, cellulose secretion, fruiting body production). The process does not stop there. The fruiting body produces spores internally. These ripen, and when they are ready, the fruiting body releases them into the environment.

Following spore dispersal, the "multicellular" pseudo-plasmodium dies, not without having contributed to the birth of a new generation of amoebas.

---

[13] An amoeba is a single-celled organism, usually with a soft body, that moves and can grab the food particles it eats thanks to the formation of pseudopods, retractable extensions of its body.

[14] The cyclic adenosine monoxide (cAMP) molecule is shown below.

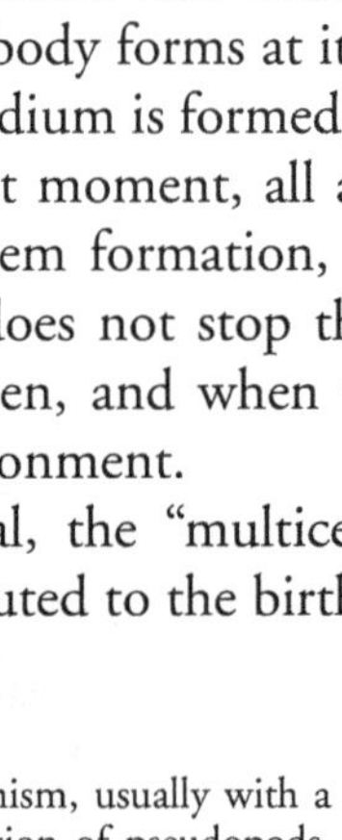

This molecule is produced from ATP by separating two phosphate groups and by interaction of the third group with the radical –OH bound to carbon in position 3' (bottom left, in the sugar ring). With two bonds, the phosphate group acquires a cyclic shape, hence the name of the molecule. Remember that ATP is the main biological energy-preserving molecule and, in eukaryotes, it is produced in the mitochondria. Tell me if you can't find a closer link between the "biological energy" and a chemical signal to communicate with your fellow man!

When a spore reaches a favourable environment, it germinates, and a small amoeba emerges to resume solitary life as long as conditions allow. As soon as these conditions deteriorate, the amoeba begins releasing cAMP again, and the cycle restarts with the formation of a new pseudo-plasmodium, stem, and spores. The process culminates with another generation of *Dictyostelium*.

Interestingly, within the myxomycetes,[15] other forms exhibit similar behaviour (for example, amoebas of the genus *Echinostelium*). The cycles differ in details (e.g., the shape of the pseudo-plasmodium and the stem), but overall, the basic mechanism is the same: when environmental conditions deteriorate, stress triggers the emission of a signal that induces conspecifics to react. They aggregate and, collectively, form a spore-producing structure that ensures a new generation of organisms.

The pseudo-plasmodium, formed thanks to the division of tasks, facilitates the dispersal of new individuals. Some cells ensure mobility, others form the stem (a real "launch" platform), and others produce spores. In this way, overall performance increases, and new amoebas are more likely to reach a new favourable environment than a solitary individual. All it takes is for a single spore to reach a region with suitable conditions, and the entire population benefits. The coordination between cells and the distribution of activities increases the chances of survival and allow myxomycetes to respond more effectively to environmental changes.

It has recently been discovered that the behaviour of amoebas derives from the phenomenon of altruism (Barbault 2006, p. 112–113)[16]: some individuals "sacrifice" themselves for the well-being of the community. The answer to the question posed at the beginning of the paragraph, regarding what drives single-celled individuals to unite into a higher entity, could therefore be altruism. This phenomenon is increasingly studied among animals, especially mammals. Who would have imagined finding manifestations of it even in the world of protists, the unicellular eukaryotes?

Regarding the solution chosen by amoebas to increase their performance, the underlying principle is the same as that observed during the emergence of the eukaryotic cell. In this case, the units at the first level "A" (prokaryotic cells) associate to form a higher-level unit "B" (the eukaryotic cell). Subsequently, "B" level units (some eukaryotic cells) associate in turn to form a new higher-level entity "C" (the multicellular organism). This modularity will be reused throughout evolution to produce further innovations.

[15] See footnote 12 of this paragraph.

[16] Chapter 6 is also a good reading to understand the phenomenon of altruism (behaviour characterized by acts that apparently do not bring any benefit to its author but are beneficial to other individuals) in living beings and the behaviours that result from them.

Let us proceed in order and return to the cycle of *Dictyostelium discoideum*. Leaving altruism aside, the real mechanism enabling the process is the specific signals (in this case, the flow of cAMP) exchanged between individuals, allowing the coordination of all cells. These signals can be produced thanks to the surplus energy of the amoeba. This surplus, ensured by the efficiency of mitochondrial action, can be diverted from cell growth and division to other purposes, in this case, to ensure chemical communication between individuals.

Once "communication" is established, coordination begins, followed by the specialization of different cells, just as at a lower level we witnessed the specialization of cellular compartments, the organelles. The emergence of this ability to communicate and behave coherently seems to coincide with the appearance of the eukaryotic cell. Although some prokaryotic associations have been observed, none appear to exhibit a comparable level of coordination (see also Sect. "Bacteria that Associate," in Appendix).

As evolution progresses, specialization increases, and the different cells making up the higher-order entity are modified to produce organs and tissues. Each element retains the same genetic heritage, but each cell (or group of cells) expresses it differently. This is how the diverse parts of a multicellular body may have evolved.

Once multicellularity has formed, another question arises: how many times has this process occurred? How many times have eukaryotic cells associated to form a multicellular organism? Once, or multiple times?

In the first case, all multicellular life (plants, fungi, and animals) would descend from the first organism that undertook this evolutionary path. In reality, the phenomenon occurred independently several times; at least seven. The process took place separately for animals, fungi, red algae, brown algae, and myxomycetes, and for green plants (green algae and higher plants) at least twice. The repeated emergence of multicellularity indicates the structural and physiological capacity of the eukaryotic cell to carry out this process.

But what are the advantages of being multicellular compared to an isolated eukaryotic cell? First, there is the ability to reach a much larger size than possible for a solitary cell. For instance, compare the size of plants and animals with that of an ostrich or dinosaur egg. The specialization of organs and tissues allows the creation of new organisms capable of exploiting resources in entirely new ways. For example, manipulating food via sophisticated mouthparts, improving movement efficiency, or developing tissues to store resources for lean times.

These are but special examples, but hundreds more exist; it is enough to observe the variety of multicellular beings that surround us: plants, fungi and animals. Despite the immense possibilities offered, some microscopic prokaryotes remain as successful within their ecological niches. Multicellularity does not necessarily produce "better" beings; it produces different beings capable of exploiting new ecological opportunities previously unavailable.

The performance of the eukaryotic cell is not limited to anatomical growth. An equally important innovation is in reproductive strategies. One of the fundamental characteristics of living beings is their ability to reproduce, producing offspring that allow the species to persist over time.

However, as noted in the first chapter, fire can also "reproduce". Do you remember that? But we know that the ways in which fire "reproduces" do not correspond to those of living organisms. Living beings, let us not forget, are chemical systems with a complex enzymatic apparatus unknown outside the domain of life.

But why do living organisms reproduce? I have no definitive answer to this question. It is one of the fundamental questions in biology, and no doubt we will find the answer one day (however, take a look at Sect. "Why Does Cell Division Exist in Bacteria?", in Appendix). For now, let us simply acknowledge that reproduction exists and that, without it, life would not have reached us. Or rather, we would not be here at this moment to appreciate it.

Consequently, we must reverse the terms of the question and ask instead: how do living beings reproduce? Consider prokaryotes, the single-celled organisms that dominated the longest interval in our planet's history. Bacteria and archaea reproduce by cell division. The so-called "mother" cell, A, grows until it reaches a certain size: the two ends move away from each other while a constriction forms in the middle, which gradually deepens until complete separation occurs. At the end of the process, there are two "daughter" cells: A' and A".

And the mother cell? It simply disappears: half of its cellular components pass to A', and the other half to A". Before division, the A cell duplicates all its enzymatic and genetic material so that each daughter inherits a full copy.

Cell division produces cells identical to the original one. The genetic heritage is identical to that of the parent, unless random variations (the so called "mutations") occur in the DNA sequence.

Sexuality, on the other hand, completely changes the process of reproduction. Sexuality is typical of eukaryotic organisms and unknown among prokaryotes. It involves reshuffling genetic material between two individuals, producing a third genetic combination: the offspring, which has an original genome.[17] This genome derives from that of the parents but is different from both, making the descendant genetically unique.

The equation that describes sexuality is: 1 + 1 → 1 (two cells fuse—e.g., an ovum, egg cell, and a spermatozoon—to produce a third element, the fertilized egg cell, or zygote), which is completely different from the cell division: 1 → 1 + 1 (a cell divides into two daughter cells).

Division separates entities; sexuality fuses them to produce a new one.[18] Given all these differences, how could the transition from cell division to sexuality have occurred?

Unfortunately, no current examples exist that show the intermediate stages. However, since it involves the fusion of two cells of the same species (in the case of the ovum and spermatozoon, the latter physically penetrates the former), it is possible that it began as a form of "cannibalism": one cell literally "ate" another of the same species. Only later would the cells have learned to pool their genetic material to produce an individual with a unique genetic fingerprint. This is when sexuality likely arose.

Of course, this remains speculative. However, the fact that sexuality differs fundamentally from cell division is attested by the necessity of a new mechanism for distributing genetic material: meiosis. Meiosis doubles the amount of DNA in a cell, which is then divided into four distinct cells. If the initial cell has a genome Y, each resulting cell ends up with Y/2 (the initial quantity doubles; then, it is divided into four different cells: Y + Y = 2Y; 2Y: 4 = 4 times Y/2).

Why would meiosis be necessary for sexuality? Imagine two cells of the same species fusing and pooling their genetic material. The resulting organism would therefore have an amount of DNA that corresponds to twice that of each of its relatives. When this, in turn, joins another cell, the DNA of the descendant will increase again. And so on and so on! To avoid this exponential explosion, a mechanism is needed to put things back in their place. In this case, we need a "trick" capable of restoring the correct amount of genetic material expected for our species. And that's exactly what meiosis does.

[17] Sexuality does not exist among prokaryotes. However, some bacterial cells can transmit a certain amount of their DNA to other individuals of the same species. This process is called "bacterial conjugation". However, it does not produce any third cell with the original genome. Simply, the DNA received finds its place in the cytoplasm of the second cell (Watson et al. 1987). The case seems not to be isolated; there are, in fact, other mechanisms that allow genetic material to travel between organisms, whether they are prokaryotic or eukaryotic (Bapteste 2013; Génermont 2014).

[18] In paragraph 2.4, I presented the idea of Patrick Forterre, according to which viruses would be real living beings, who have acquired a very particular way to produce descendants, completely different from that of prokaryotic cells or from that of eukaryotes. Viruses pass from an acellular phase, the virion, which must necessarily infect a cell (prokaryotic or eukaryotic) in order to perpetuate the virus's descent. In the current paragraph, I focus only on the reproductive modes of bacteria and archaea, on the one hand, to that of eukaryotic organisms, on the other.

In sexually reproducing organisms, there is an alternation of generations: individuals with genetic heritage X unite to produce generation 2X. These 2X individuals, through meiosis, produce X individuals again, and the cycle continues. In fact, there are no crosses between the odd and even generation (between X and 2X individuals).

Let us take the case of plants. A tree, for example, is an individual produced by the union of two sex cells. It is part of the 2X generation which is called "sporophyte" because, thanks to meiosis, it produces spores. These are part of generation X: they are the "gametophytes" that will produce gametes, sex cells. Generation X, in many plants, is small in size and lives at the expense of generation 2X. In angiosperms (flowering plants) the gametophytes include the pollen and the embryonic sac (which produces the oosphere) found in the flowers. But in different vegetables, for example, mosses, the situation is reversed: the gametophytes form the visible part of the vegetable. Sporophytes (the 2X individuals), by contrary, are hardly visible; they live hidden from the individuals of the other generation, on whom they depend.

In humans, individuals are diploid, with body cells containing two copies of each chromosome[19] (our genetic heritage). Our sex cells (spermatozoa, in men, and ova, in women[20]), are haploid, with only one copy of DNA. They represent the generation X. In our case, as in that of other animals, generation X does not possess a reality independent of the individuals who produce it.

Sexuality, and meiosis in particular, requires specific genes. Remarkably, a protein necessary for the realization of the meiotic process (therefore, indispensable for sexuality) has an enzymatic equivalent in a hyperthermophilic archaeon (Forterre 2007), that does not know sexuality. This example highlights the phylogenetic connections between the two domains: eukaryotic cells themselves may have arisen from a symbiosis between bacteria and at least one archaeon, according to the theory developed in the previous section.

Despite its mysterious origins, sexuality exists and has imposed itself among eukaryotes. But why? What are their real advantages? If we examine the process of meiosis and fertilization (the fusion of two sex cells, the gametes), we realize that they create extensive reshuffling.

In humans, each of us inherits roughly half of our chromosomes from each parent. This realizes the first reshuffle: the genes of an individual do not

---

[19] Chromosomes, in sexually endowed organisms, are the units of DNA segregation.

[20] In our species, as in many others, the sexes are separated; that is, there are male individuals producing spermatozoa and female individuals producing egg cells. However, both among the animals (the Burgundy snail, to give an example) and among the plants, we find species in which the same individual produces at the same time, spermatozoa and ova. Sexuality simply needs the interaction of two individuals to produce descendants; does not necessarily require separation of the sexes.

simply correspond to those of the father or mother, but derive from a genetic mixture of the two.

But there is more; if we examine in detail the phenomenon of meiosis (which produced paternal spermatozoa and maternal egg cells), we discover that, thanks to chromosomal segregation and the phenomenon of crossing-over,[21] the possibilities of genetic combination are enormous. This means that the hundreds of millions of egg cells (or spermatozoa) that can be produced by the same individual are ALL different. Even two siblings (except identical twins) are genetically distinct because they are produced from different gametes.[22]

The advantage of all this genetic "upheaval" from one individual to another provides first-class material for biological evolution. Natural selection acts on variation; when all individuals differ genetically, selection has much material to work with, unlike prokaryotes, where evolutionary novelty depends on rare mutations. In bacteria, for example, mutation rates are roughly 3 mutations per 1 000 generation.[23] Since most of these are harmful to the body, the simple mechanism of random mutation places precise time constraints on the speed of evolution. Despite the fact that in the world of prokaryotes generational renewal can take place every 20–30 min (e.g., a cell of *Escherichia coli*, a bacterium living, among other things, in our intestine, can divide after only 20 min from its formation) the times to allow the appearance of an evolutionary novelty remain relatively long. Instead, thanks to the genetic mixing produced by sexuality (due to three different processes: fertilization, chromosomal segregation and crossing-over) all descendants are different.

Mind you: sexuality does not create new genes. It simply shuffles those that are already present in the population. As a result, it creates variety, a lot of variety! It is as if evolution came down from the bicycle, he used to ride with prokaryotes to take a Ferrari. As André Langaney notes (1987,

---

[21] Crossing-over is a process of gene recombination (that is, a phenomenon which allows the arrangement of an individual's or cell's DNA according to a different association than that observed in cells or parental individuals) which involves genes within chromosomes and occurs during one of the stages of meiosis (Watson et al. 1987).

[22] Meiosis transmits only half of the complete genetic pool in each gamete, which consists, I recall, of two copies of each chromosome. In the man, who has 23 chromosomes, the number of possible spermatozoa is, thanks to chromosome segregation 2^{23} (more than eight million), which is also equal to the number of possible ova produced by the other sex. At the time of fertilization, the possible combinations resulting will therefore be $2^{23} \times 2^{23}$, a number equivalent to several thousand billion. And we have not even considered the countless possibilities of crossing-over! (Langueney 1987, p. 23–24).

[23] The random mutation rate is, in bacteria, of the order of 5×10^{7} per gene and par generation. The result reported in the text was derived in this way (Ninio 1996). However, to fully appreciate the value of evolution in bacteria, I also recommend reading the text by Pennisi (2013).

p. 29), "Sexuality accelerates to the extreme the otherwise very slow pace of evolution."[24]

I would end this "overview" dedicated to sexuality by pointing out, once again, that prokaryotic organisms ARE NOT at all less evolved than eukaryotes. Sexuality increases genetic variety and differentiation (and also morphological differentiation), but the *Escherichia coli* cells, which live in our intestines, are just as evolved as whales or the latest Nobel Prize in physics! All of them are the product of more than 3 billion years of biological evolution. Sexuality and multicellularity do not produce "better" organisms; they produce organisms capable of exploiting new ecological niches and solving different biological challenges.

3.3 The Oldest Traces of Eukaryotes

When did eukaryotes appear? What is their age? What are their oldest fossils? Similar questions were already raised in paragraph 2.1 regarding the oldest traces of Life on Earth. Here, we revisit them, focusing on organisms of a particular domain: nucleated cells, the eukaryotes.

Unfortunately, the nucleus does not fossilize. To identify these organisms, researchers must rely on other criteria, one of which is size. This is why some Archaean fossils continue to intrigue specialists. The spherical cells discovered in the Moodies Group sediments (South Africa), dated to more than 3 billion years ago and are among these. These cells are exceptionally large, reaching a diameter of almost 300 µm (Javaux et al. 2001). Despite the fact that some current prokaryotes, spirochetes, possess cells 500 µm long, same specialists issued the hypothesis that the Moodies Group vestiges may have belonged to eukaryotes. Is this assumption plausible? Could eukaryotes be this old? While proponents of the strict tripartition of life see no objection, the fact remains that, after this period, fossils of potential nucleated cells remain unknown for a long time. Apart from size, no other evidence allows us to attribute a eukaryotic nature to these African fossils.

To find other traces of presumed eukaryotes, one must move forward to around 2.7 Gy. It is precisely in the strata of this era that chemical evidence indicates the presence of nucleated cells in the late Archean. An Australian team (Brocks et al. 1999) identified chemical components characteristic of eukaryotic cells in sediments from two Australian formations: the

[24] This work is indispensable to appreciate the innovations due to sexuality! Those who are comfortable with English can read Chap. 5 of the book by Nick Lane (Lane 2010) for good. Regarding the evolution of eukaryotic sexuality and its importance see also Génermont (2014).

Marra Mamba Formation, about 2.6 Gy old, and the slightly older Maddina Formation (~2.7 Ga).

What did the Australian researchers find exactly? They detected traces of steranes in the hydrocarbons isolated from these sediments. Steranes are molecules consisting of three six-carbon rings plus a fourth five-carbon ring. Sterols, which feature four condensed rings with a hydroxyl group (–OH) at the 3' carbon, belong to this family. Beyond the chemical specifics, the key point is that sterols are typical of eukaryotic cells, where they play an essential physiological role; cholesterol is the most famous example.[25]

This discovery allows us to affirm that, at least from the terminal phase of the Archean (~2.6 Gy), representatives of the three domains of life, bacteria, archaea, and eukaryotes, could have been present on Earth.

During this period, the only detectable traces are chemical. More complete eukaryotic fossils appear in the first half of the Proterozoic, around 2.1–2 Gy. In these strata, researchers have excavated eukaryotic fossils, some measuring up to several centimetres.

The oldest remains, dated to 2.1 Gy, come from Gabonese sediments. These fossils include structures ranging from 7 to 120 mm, interpreted as vestiges of colonial organisms. Chemical analysis indicates the fossils were deposited beneath a layer of hydrogen peroxide. Morphological studies revealed similarities to the enigmatic organism *Mawsonites spriggi*, a medusoid species whose phylogenetic relationships remain unresolved, known from younger layers several hundred million years old.

While some modern prokaryotes can form large clusters (up to 15 cm), several internal structural features of the Gabonese fossils suggest genuine eukaryotic organisms. Even the apparent construction of "colonies" follows more complex growth rules than those observed in prokaryotic aggregations. All of this suggests they were communities of eukaryotic cells (El Albani et al. 2010). However, in this context, interpretations remain tentative due to the lack of more definitive data. Even their assignment to eukaryotes could be questioned based on other evidence.

Another problematic fossil is that of *Grypania spiralis*, with older remains (dated to 2.1 Gy; Han and Runnegar 1992) contemporary with the Gabonese vestiges. This organism appears to have been relatively successful, as its remains are relatively abundant even between 1.8 and 1.4 Gy (Porter 2004). What did *Grypania* look like? Its shape is distinctive: an elongated, spaghetti-like mass, without branching or dichotomies (Fig. 3.4). But what is it really? A single-celled organism with unusual morphology? Or a colonial organism?

---

[25] Although sterols are typically eukaryotic, some bacteria can synthesize cholesterol. However, they do so using genes acquired from eukaryotes via lateral gene transfer (Knoll 2004, p. 94, footnote).

Despite the abundance of fossils, the question remains unresolved. Specialists agree only on one point: it is a true eukaryotic organism, likely a type of unicellular algae.

The interpretation of *Bangiomorpha pubescens* (Fig. 3.4) is, fortunately, clearer. The fossil was discovered in the Canadian Arctic in layers aproximatively 1.2 Ga old (Gibson et al. 2018). Detailed study of the specimens allowed researchers to establish a "portrait" of the organism, showing a close relationship with red algae, specifically the modern genus *Bangia*. The fossil's generic name, *Bangiomorpha*, is therefore not accidental (Butterfield 2000).

This discovery not only confirms that red algae existed in the Proterozoic at least as a group, but also marks a major biological milestone. The genus *Bangia*, like other algae in the same group, is a multicellular organism capable of sexual reproduction. This implies that, at least 1.2 billion years ago, two key innovations of eukaryotic cells, the ability to form multicellular individuals with differentiated cells and the emergence of sexuality, had already been achieved!

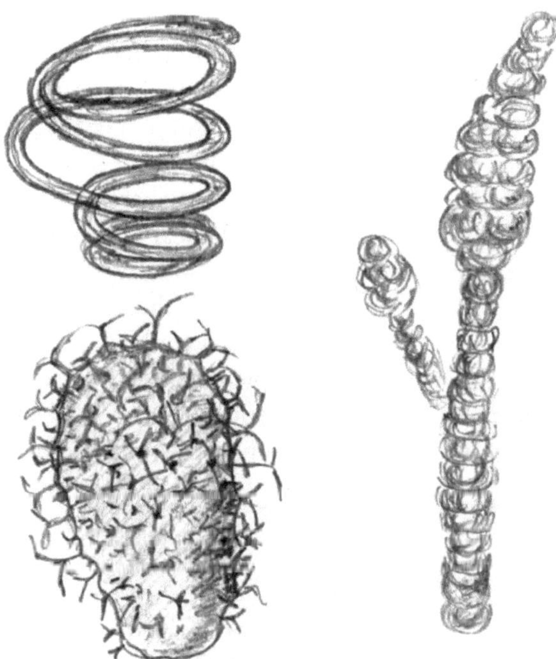

Fig. 3.4 Upper left: fossil of *Grypania spiralis* (the illustrated organism is approximately the size of a ten-cent coin); right: reconstruction of *Bangiomorpha pubescens* (the specimen's height is about 200 μm, i.e., 0.2 mm), while the filament of the algae could reach 800 μm (just under one millimetre); lower left: a fossil from *Tappania* (height about 300 μm, i.e., 0.3 mm)

Moving on to other eukaryotic fossils from roughly the same period (~1 Ga), we encounter *Palaeovaucheria clavata*. This fossil may have possessed "secondary" chloroplasts (organelles derived not from prokaryotes but from eukaryotic photosynthetic cells), similar to modern algae of the genus *Vaucheria*. In these ancestral algae, the host cell did not form a symbiotic relationship with a cyanobacteria, but directly with a eukaryotic photosynthetic organism, which later became the secondary chloroplast.[26] This mechanism mirrors that proposed by Lynn Margulis for the formation of the eukaryotic nucleus, but here it involves two eukaryotic cells: one becoming the host, the other the photosynthetic centre.

And what about *Tappania* (Fig. 3.4)? Some researchers have interpreted it as a fungus (Porter 2004)[27]; if correct, this would make it the oldest known representative of this group. From 1.5 Ga onward, one of today's main eukaryotic groups (fungi) may already have been present!

In addition to this, starting from this period, the fossil evidence attests to the presence of entire communities of protists (unicellular eukaryotic organisms) capable of highly differentiated ecological structures.[28]

No account of Proterozoic eukaryotic fossils is complete without mentioning acritarhs (Fig. 3.5). What are acritarchs? They are small organisms with organic walls that are insoluble to acids and cannot be classified into another known group. When researchers encounter a microfossil of uncertain attribution, they classify it as an acritarch if it meets these characteristics. This group is heterogeneous, but all members share resistance to the acids commonly used in geological and paleontological preparation.

Occasionally, specimens are re-evaluated and may change systematic classification. Taxonomy can vary depending on the specialist examining the organism. This applies to *Tappania*, initially described as a "mushroom" by its discoverer, but later considered as a simple acritarch (Huntley et al. 2006).[29] Acritarchs have been interpreted as particular algae, specialized unicellular organisms, or even spores. Most authors, however, classify them as eukaryotes.

---

[26] A secondary chloroplast is an organelle which does not derive from a cyanobacterium, but from a eukaryotic photosynthetic organism, a nucleus cell itself, provided with chloroplast. For more information, see Chap. 9 of the book by Selosse (2017).

[27] However, the idea is now beginning to develop that the ancient forms of *Tappania* are clearly different from the younger ones (Porter 2006). In this case, only the latter (found in sediments between 900 and 800 Ma), would be true "fungi" fossil. For other possible fossil fungi, see Loron et al. (2019).

[28] The paper by Javaux et al. (2001) describes fossil remains found in the Ruper Group (northern part of Australia), at internodal layers dated to about 1.5 Gy. The picture that emerges is that of a marine environment inhabited by eukaryotic organisms very sophisticated both from an anatomical point of view and ecological organization. Regarding the diversity of eukaryotic organisms in the Proterozoic, interested readers may wish to consult Knoll et al. (2006).

[29] This paper is the source of practically all the information in the text about the acritarchs.

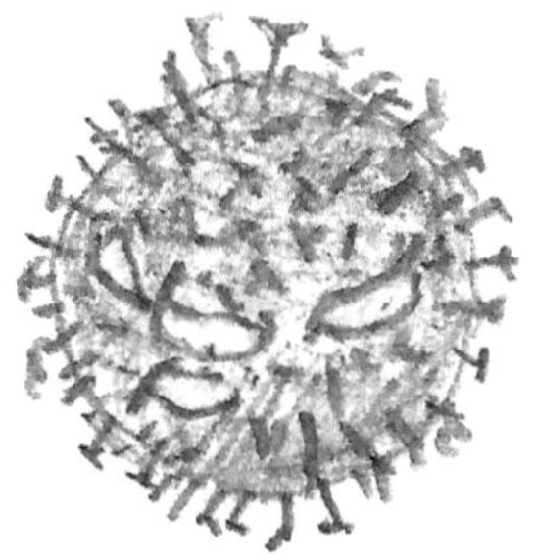 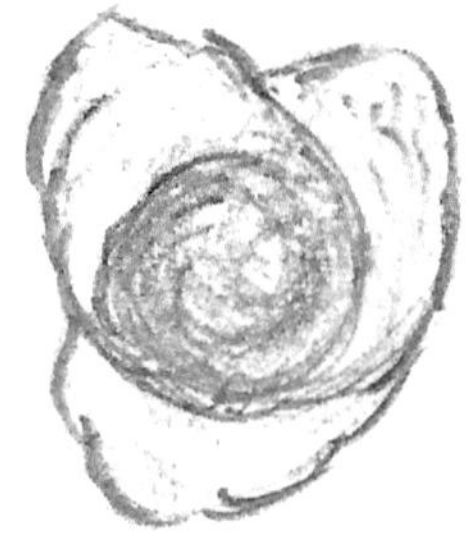 

Fig. 3.5 Three acritarchs. Two, on the left and in the middle, came from the Ruyang group (China), between 1.7 and 1.4 Gy in age; the last, on right, is more recent (Ordovician, ~ 490 My) and come from Estonia. The three fossils are tiny and they are represented at different scales

Their significance lies in their longevity: the oldest traces date to the Proterozoic (1.8–1.6 Ga), while the most recent are 488 My old, reaching the end of the Cambrian. Due to their diversification and long existence, these organisms are useful biochronologically; identifying specific taxa allows Proterozoic specialists to date fossil layers without other methods.

Acritarchs seem to have endured several environmental and climatic changes. Furthermore, their diversity increased at the beginning of the Cambrian (~540 My ago), coinciding with one of the greatest animal diversifications, suggesting a possible ecological or tropic link.

Finally, we end our excursion on Proterozoic biological communities by reporting one last fact: the oldest known terrestrial organisms also date from this period. Fossils indicate that by 1.2–1 Ga, some organisms had left the sea, including eukaryotes, prokaryotes, and cyanobacteria-related organisms (Strother et al. 2011; Horodyski and Knauth 1994). But how did these creatures avoid harmful space radiation that still reached Earth's surface? These pioneers likely sought shelter in protected environments such as karst environments or fresh waters (rivers, lakes, estuaries, or simple bodies of water, more or less extensive). And yes… once Life has established itself, nothing can stop its action!

Above, I recalled the prokaryotes. What else can be said in their favour? Although eukaryotes were diversifying, prokaryotes remained an important component of biodiversity throughout the Proterozoic. For example, between 1.3 and 1.2 Ga, a peak in stromatolite diversification is recorded (Taylor et al. 2009, Fig. 2.35 p. 67; For information on stromatolite, see paragraph 2.1). Numerous paleontological sites are renowned for their prokaryotic fossils (Knoll 2004, Chap. 6). For instance, the Bitter Spring Formation, located in central Australia, contains fossil layers 830–800 Ma old, preserving a rich flora of microorganisms identified as cyanobacteria. In contrast, the older layers of the Gunflint Formation (Ontario, Canada), about 1.878 Ga, preserved other types of prokaryotic organisms. Although associated with stromatolitic structures, comparative analyses with modern microbes indicate that they were not photosynthetic bacteria. Instead, they resembled microorganisms of the genera *Leptothrix* et *Sphaerotilus*, sulphur-reducing prokaryotes inhabiting iron-rich, oxygenated waters and possessing a metabolism very different from cyanobacteria. Terrestrial prokaryotes, such as those discovered in Arizona, are further evidence of their widespread distribution (Horodyski and Knauth 1994).

Fossil evidence and modern data demonstrate that prokaryotes were not negatively impacted by the emergence of eukaryotes. Instead, they continued to evolve independently, thriving in their specific ecological niches. Since their appearance between 3.8 and 3.5 Ga, bacteria and archaea have contributed significantly to global biodiversity and the network of biological interactions among living organisms.

3.4 Precambrian Climate Changes

We have witnessed the emergence and evolution of organisms from all three domains of life. Paleontological data indicate that certain forms were typical of specific periods, while others persisted over long intervals. Why do these changes occur? Why do living organisms not retain the characteristics of their earliest forms?

Organisms that remain unchanged can only survive in stable environments. In reality, Earth has undergone constant transformations since its formation, altering the habitats it hosts. Consequently, organisms must adapt to new conditions or face extinction: this is the fundamental principle of biological evolution.

Now, we focus on some of the changes occurring during the Precambrian[30] as they are closely linked to the emergence and evolution of life.

These changes can be categorized into two types: biological and physical, which mutually influence each other. Biological changes involve the activities of organisms modifying their environments, for example, by exploiting

[30] The Precambrian is the interval of time that goes from the formation of the Earth, placed around 4.6 Gy at the beginning of the Palaeozoic, about 541 My. Practically, it includes all the time period we have considered so far. Apart from the Adean (between 4.6 and 4 Gy, which I have neglected because, as far as we know, it is totally sterile), the two eons of which we have dealt with, the Archean and the Proterozoic, constitute the two major divisions of the Precambrian.

resources or generating and accumulating waste. A particularly important example is the early production of oxygen by photosynthetic organisms.

Among the earliest known life forms are microorganisms resembling modern cyanobacteria.[31] They shared the same metabolism, capable of photosynthesis similar to that of green plants, producing oxygen as a by-product. During the Archean, these cyanobacteria released oxygen into aquatic environments that initially contained none.

Before continuing, it is important to recall a key aspect of the Archean environment: the primordial atmosphere lacked free oxygen. The transition from non-living to living systems could not have occurred in the presence of this reactive gas, which would have hindered the chemical evolution required for life. Today, oxygen is abundant, and spontaneous emergence of life is no longer possible; life now only arises from pre-existing organisms. In the early Earth, however, conditions were quite different: they allowed the first organisms to emerge and evolve with the passage of time.

Among the current ideas on this topic, I have given particular importance to those put forward by Professor Mario Ageno. He emphasized the crucial role of photosynthesis in the origin of life. Fossil evidence supports this, with "cyanobacterial" vestiges found in 3.5 Ga sediments, indicating that oxygen production began very early and has continued ever since.

Nowadays, perhaps, it is difficult to realize what the rejection of this gas could have implied in an environment completely devoid of it. I will try to illustrate it with an example.

Imagine a company producing a unique drink that satisfies all dietary needs. Everyone consumes it daily, and the containers are discarded without concern. Nobody cares about their disposal: people have other things on their minds! Over time, landfills overflow, eventually encroaching on living spaces. Only then do people start realize that their environments are in danger, but only a few become able to escape the scourge or turn the situation to their advantage.

Some might think that the metaphor is a satirical representation of the mass consumer society that risks being submerged by its waste. In fact, this is the situation that every living community (be it humans, animals, or whatever) faces with respect to its waste if it is not recycled in some way. What is produced with oxygen is no exception. This scenario metaphorically represents the accumulation of such a gas produced by photosynthesizing organisms. In a previously anoxic environment, oxygen was effectively a

[31] I have already introduced these beautiful organisms, paragraph 2.2. Do you remember it?

"waste product" with significant ecological consequences, representing the first biologically-driven environmental change on Earth.

But what evidence do we have of these events? The most important come from sedimentary rocks.[32] Around 2.4 Ga, a major transition occurred. Before this time, oxygen released by cyanobacteria remained diluted in the oceans, leaving little to no trace in sediments. Minerals such as pyrite, siderite ($FeCO_3$), or uraninite (UO_2) were deposited without oxidation, because this gas was too diluted in the environment. After 2.4 Ga and up to ~ 1.8 Ga, sedimentary layers display iron deposits that appear red; essentially "rust" from the reaction of iron ions with oxygen. These "red beds" or "iron beds" provide a record of increasing environmental oxygen levels.[33] Some specialists suggest that by 2.2 Ga, atmospheric oxygen may have reached ~ 1% of the current concentration.[34] Regardless of the exact timing, the formation of red beds confirms the accumulation and diffusion of oxygen in the Precambrian environment.

And what were the consequences of the presence of oxygen in the environment? Apart from the purely sedimentological effects, this gas certainly influenced the communities of organisms. Its presence in the environment has meant the oxidation of various elements or molecules, which have therefore been altered from their initial state.[35] In the case of the key resources of

---

[32] The main facts about the evolution of oxygen concentration in the environment are detailed in the book by Knoll (2004, Chap. 6).

[33] When gaseous oxygen is rare, iron comes in the form of ferrous ions. Being soluble, these ions are carried by currents and generally do not form important accumulations in the sediments. When the oxygen content rises above a certain threshold, iron is transformed into insoluble, oxidised ions which settle to form red viscous layers. These are the "red-beds" or "iron-beds", typical of rocks aged between 2.4 and 1.84 Gy, which testify to the oxygen content in the environment. Note that these deposits have the typical appearance of stromatolitic layers, which testifies, once again, to the close link between living beings and oxygen emissions. Finally, I would point out that this first peak of oxygen seems to coincide with the first confirmed fossils of eukaryotic organisms. About the relationship between these two phenomena, see Mieli et al. (2025).

[34] Of course, such assessments are difficult to make and it is possible that the reported content will need to be corrected in the future. For example, Fedonkin (2003) reports that at the same time, the oxygen content could have been much higher. Currently, this gas is present in the Earth's atmosphere at about 21%. Nitrogen is the most common gas (~78%). Carbon dioxide (CO_2), is present only in a small percentage (~0,023%).

[35] Towards the end of the Archean and the beginning of the Proterozoic, the main carbon cycle used by living beings on the seabed was different from that used by modern organisms inhabiting the same environment. Nowadays, under the sediments, oxygen is rapidly consumed and can live in these environments are organisms with anaerobic metabolism. The most abundant microbes in this environment are currently those capable of reducing sulphates. This operation requires the presence of oxidized sulphates which can be reduced. However, the absence of oxygen at that time prevented the formation of these compounds. The main carbon cycle was not based on sulphate reduction but on methane production (the main inhabitants of the ocean floor were probably methanogen archaea, like those living today on the bottom of anoxic lakes). It was the production and diffusion of oxygen that allowed the oxidation of sulphates and, over time, the establishment in marine sediments of a metabolism based on the reduction of these molecules.

certain organisms, they were faced with a dilemma: either they would become able to find others to exploit or they would disappear.

But things were perhaps even more complicated. Oxygen was also toxic to some microorganisms, such as methanogenic archaea, which can only survive in anoxic environments; we even find them in our intestines. In this particular environment, the presence of other aerobic microorganisms (which thrive in the presence of oxygen) is enough to consume all the gas and allow methanogenic microbes to develop. But if the former were to disappear, the methanogens would soon perish.

The diffusion of oxygen represented, without a doubt, a scourge of the first order for all those organisms that lived in the anoxic environments of the Archean. The spread of oxygen likely caused a wave of extinctions among organisms that could not tolerate it or whose resources were compromised. Some organisms found refuges where oxygen could not penetrate, but many disappeared. This diffusion of oxygen represents one of the first biological crises caused entirely by the metabolic activity of living organisms.

However, such crises also create opportunities. Following the extinction events, organisms that could tolerate oxygen filled the ecological niches and gradually became dominant.[36] The release of oxygen gradually transformed the atmosphere into its current composition. In short, the first photosynthesizing organisms "polluted" their environment with oxygen, laying the foundation for a planet habitable by animals, including humans!

Not all changes were biological. Some were purely physical. Around 1.2–1.0 Ga, continental landmasses began to unite, forming the first supercontinent, Rodinia (Rogers 1996). Rodinia did not remain intact indefinitely; by around 700 Ma, it began to fragment. Continental movements (due to the phenomenon of continental drift) influenced coastal environments, which are among the richest in biodiversity underwater.[37] When landmasses consolidate, these environments shrink, reducing global biodiversity. Conversely, when continents break apart, coastal zones expand, and associated organisms proliferate.

Climate change also impacted biodiversity. During the Proterozoic, glaciations of varying duration and extent occurred. In some periods, ice may have reached tropical regions, explaining glacial deposits at low latitudes. Some

[36] The Gabonese fossils, dated at about 2.1 Gy (I have already mentioned it in the previous paragraph) could be an example of organisms that have prospered thanks to the first phase of oxygenation of the aquatic environment, which began around 2.4 Gy. The sediments in which the fossils have been preserved are evidence of a low-depth, oxygenated aquatic environment (El Albani et al. 2010).

[37] About 80% of the biomass of benthic marine organisms occupies coastal environments (between 0 and 200 m deep, while this type of environment constitutes less than 8% of the ocean floor (Fedonkin 2003).

scientists suggest Earth may have been completely ice-covered at times (the "Snowball Earth" hypothesis), both towards the beginning of the Proterozoic, around 2.3 and 2.2 Gy, and in its final part, around 750 and 580 My (Hoffman et al. 1998; Ashkenazy et al. 2013; Sansjofre and Hir 2003).

Is such a scenario possible? Is it possible that Earth has turned into a kind of "ice planet" just like Mars? And if it was possible, how did he return to his original condition?

In this regard, scientists still debate, with supporters of the full ice extent on the Earth's surface (land and oceans), and others believing that ice-free regions existed, where photosynthesis of primary producers (algae and cyanobacteria) may have taken place.

Regardless of the details, Precambrian glaciations ended, and life rebounded. Most major evolutionary lineages of prokaryotes (bacteria and archaea) and eukaryotes (ciliates, various types of algae, fungi, etc.) survived these adversities. After the final glacial cycles of the Proterozoic, the first traces of animals appear (Huntley et al. 2006). Could these events be connected?

3.5 The Emergence of Animals

Paleontological data indicate that some of the main representatives (or their precursors) of the major multicellular eukaryotic groups (green, brown and red algae [Taylor et al. 2009] and fungi) were already present. But we have not yet discussed animals. When do animals appear?

First, let us consider what are the characteristics that make an animal a… animal. Without providing a strict taxonomic definition, I define animals based on the following criteria:

- they are multicellular eukaryotes with differentiated cells, forming tissues and organs with specialized functions;
- they are heterotrophic, meaning they cannot produce their own food and must obtain nutrients from their environment;
- they possess the ability to move, at least during one or more stages of their life cycle;
- during their development, they pass through an embryonic phase with specific stages.[38]

[38] Fungi form another group of heterotrophic eukaryotes, but are characterized by cells containing chitin. In addition, mushrooms feed by "absorption" (they make the nutritional elements penetrate into their interior—body or cell) while animals feed by "ingestion" (food enters through an orifice—the mouth—into the digestive system). Most taxa belonging to fungi reproduce themselves

The first traces of animals appear to have been identified in the Doushantuo Formation (China), dated to approximately 600 Ma (Porter 2004). These layers preserve a rich community, including microfossils of prokaryotic organisms, red and brown algae, numerous acritarchs, ciliated protists (a type of unicellular eukaryote), and even possible cnidarians (the group that includes jellyfish and corals), as well as sponges. The most remarkable fossils, however, are those interpreted as possible animal embryos.[39] Despite the efforts made by specialists, the belonging of these embryos to a specific group of animals has not yet been established. However, this does not detract from the interest of discovery.

Genuine remains of sponges have been found in the Doushamatuo formation. Sponges, or Porifera, are real animals, although their anatomy is very simple. This does not detract from the fact that it is perfectly functional for their lifestyle. A sponge is made up of a mass of more or less spongy tissues, despite being supported by a calcareous or silicious skeleton formed by spicules. Water is inhaled into a body cavity where it is filtered. Food particles are assimilated, while all elements without nutritional interest are excreted along with waste. Despite being simple, sponges meet all the criteria indicated to be considered as animals in their own right.[40]

As mentioned above, sponges have a skeleton of spicules, the interest of which lies in the fact that they can fossilize and be found later. This is what happened in the Doushamatuo Formation. The Chinese remains have been identified as part of the demosponges, or siliceous sponges, because they have spicules formed by this mineral. Fossils of photosynthesizing organisms closely associated with the remains of sponges have also been found (Li et al. 1998). This is an association between algae and porifera that still persists to this day.

---

through spores; embryos are unknown. For more information on animals see the digital encyclopaedia Wikipedia, at the enter "Animalia"; for fungi, see the Wikipedia entry "Fungi". For further details on the definition of animals, see Lecointre (2009).

[39] The specialists continue to debate the real identity of these last fossils. Some consider them to be true embryos (Xiao et al. 1998; Chen et al. 2004), while others argue that the structures that led to their identification are simply the result of taphonomic alterations (Bengtson and Budd 2004). Currently, the majority of paleontologists recognize some of these remains as fossil embryos. The x-ray study allowed us to evaluate the cell division stage and the affinity with certain higher taxa of animals at phylum level (Hagadorn et al. 2006).

[40] Despite the morphological simplicity of sponges, today we know even simpler animals. This is the case of *Trichoplax adhaerens*, a placozoan. It is a tiny being with a flattened body, without mouth, intestine, nervous system or even extracellular matrix. Placozoa crawl on the seafloor and live in all the warm seas of the planet. Based on cell genetics studies, specialists believe that their origins date to around 600 My. Unfortunately, no genuine fossil of such creatures, or its precursors, has ever been found. (Voigt et al. 2004; see also J. Bischoff's text in Collective Work 2013, pp. 66–67).

The symbiosis between modern sponges and photosynthesizing organisms is relatively common in the shallow waters of tropical and subtropical seas. For example, half of the species of today's Caribbean porifera form a symbiosis with cyanobacteria or algae. Chinese fossils, on the one hand, show us that this alliance is very old (~580 My), on the other hand, they remind us of the importance that associations and symbioses have for biodiversity even since ancient times.

If the previous example is not sufficient, here is another association, also identified in the Doushantuo Formation, thanks to the meticulous analysis of the fossils found there. Specialists recognized vestiges interpreted as ancient "lichens" (Yuan et al. 2005). Quotation marks are necessary here because palaeontologists highlighted a situation very similar to modern associations between fungi and algae (or between fungi and cyanobacteria), which characterizes modern lichens. However, they were unable to identify the exact organisms involved. Nevertheless, if interpreted correctly, these results show that a lichen-like association preceded the colonisation of the land by plants and fungi (it is no wonder that they are sometimes referred to as "aquatic mushrooms," since modern species exhibit such a lifestyle). Additionally, these fossils may provide further evidence for the presence of fungi in the biological communities of the time.

Biodiversity appears to increase further in the strata between 575 My and the beginning of the Cambrian, around 541 My. This period is called "Ediacaran",[41] named after an Australian locality north of Adelaide, which, at the end of the 1800s, yielded representatives of a very interesting biological community.

Later, Ediacaran organisms were found in many countries, including Australia, Russia, Namibia, Mexico, and China. The oldest representatives of these communities date to about 600 My, while the youngest have been found in Middle Cambrian strata, between 510 and 500 My. During this period, faunas and fossil floras were completely different from those of the Precambrian, and the Ediacaran organisms that survived are considered relicts of a more ancient era.

What were the characteristics of these communities? What types of organisms were present? Together with macrofossils (the actual Ediacaran organisms), several traces were identified, showing that ancient organisms carried out their activities in sediments. Remains of algae and bacteria were

---

[41] To be more precise, the Ediacarian, according to the international biochronological scale of 2019 (see the appendix, at the end of the book), begins around 635 My, but, including the age of the fossil layers of the Doushamatuo Formation mentioned in the text. The most well-known Ediacarian communities, however, are part of the final Proterozoic, after 575 My.

also found. The latter often covered the Ediacaran creatures with a biofilm that helped preserve the organic remains of the macrofossils. Most Ediacaran organisms were exhumed in sandy layers (storm deposits or debris flows), and many macrofossils even contained sand.

And macrofossils, what do they look like? Some macrofossils have a flattened, round shape, like a plate or the umbrella of a jellyfish (the enlarged upper part under which the tentacles extend). Others have a more or less rounded silhouette with ribs branching from a central groove, like *Dickinsonia* (Fig. 3.6), a vermiform appearance with a kind of cephalic shield (*Spriggina*; Fig. 3.6), or a triradiate structure on the back (*Tribrachidium*; Fig. 3.7). Still others have the appearance of large leaves, such as *Charnia masoni* or *Charniodiscus* (Fig. 3.8). The latter taxon resembles a feathered branch whose stem ends in a rounded base that anchors the organism to the substrate. The presence of sand inside the rounded structures has been interpreted as "weights" preventing the body from tipping over due to currents or other disturbances. Other images and descriptions can be found on the Wikipedia site dedicated to Ediacara faunas or in Mark McMenamin's book (1998).[42]

Beyond the particular aspects of each taxon, the characteristics that seem common to all fossils typical of this community are the lack of rigid tissues (apart from a few exceptions, discussed later, these creatures had to be covered with a sort of chitinous tissue), the absence of sense organs (such as eyes, nose, mouth) and a flattened morphology.

Not all fossils in the Ediacarian were covered with soft tissue. At least two different taxa, *Cloudina* and *Namacalathus*, possessed a mineralized outer skeleton (Knoll 2004, Chap. 10). These two taxa have been found in Namibian sediments between 550 and 543 My in age and are the oldest organisms with mineralized tissues.

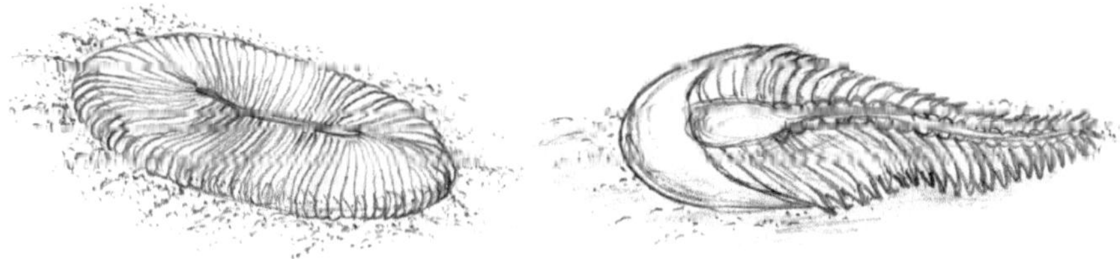

Fig. 3.6 On the left, reconstruction of *Dickinsonia costata*, an enigmatic organism, represented here on sandy soil (total length, about fifteen centimetres); on the right, reconstruction of *Spriggina* (total length, about five centimetres)

[42] This work is also the main source of information on Edacarian organisms used in this paragraph.

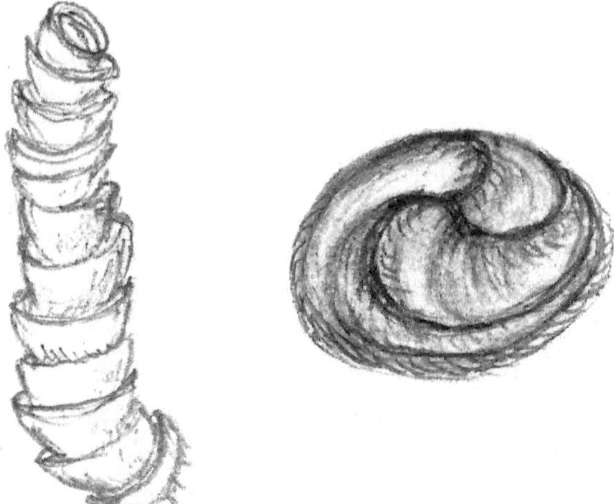

Fig. 3.7 On the left, *Cloudina*, Ediacaran organism with mineralized exoskeleton (Esmeralda County, Nevada); the whole structure is a few millimetres high. On the right, *Tribrachidium heraldicum*, a triradially symmetric fossil, about two centimetres in diameter (Australia)

Fig. 3.8 On the left, fossil of *Charnia masoni* (total length about 20 cm). On the right, reconstruction of *Charniodiscus* in life position (total height about thirty centimetres)

Claudina fossils (Fig. 3.7) are small tubes made of weakly mineralized calcium carbonate. Each resembles a stack of small bowls arranged on top of each other. The organism likely lived inside the tube. Some structures have perforations, but it is unclear whether these holes were caused by predatory activity or are simply the result of post-mortem degradation. If the first hypothesis is correct, these would represent the oldest known traces of predation. However, no trace of a hypothetical Ediacaran predator has ever been found, either in Namibia or elsewhere. Similar fossils, distinct enough to warrant a different scientific name, have been discovered in China and North America.

Namacalathus is completely different. It likely resembled a rounded cup with six or seven large openings on its surface, supported by a relatively flexible cylindrical peduncle. As in the case of *Cloudina*, the mineralized tissue was very weak, providing minimal protection. Furthermore, no predatory organisms have ever been found in Precambrian formations.

After examining the Ediacaran creatures, let us now consider the interpretations that have been proposed to understand their phylogenetic relationships with other organisms, both extant and fossil.

The first interpretation was to regard Ediacarian organisms as precursors of the systematic groups of modern or Palaeozoic animals (Erwin and Valentine 2013). For example, *Spirggina*, with its elongated body equipped with lateral lobes and a cephalic shield (see Fig. 3.6) has been interpreted as a possible arthropod precursor of trilobites (animals that would develop only later in the Palaeozoic seas). Circular shapes have been interpreted as jellyfish fossils, while the taxa *Charnia* and *Charniodiscus* were considered particular algae or even pennatulacea, colonial organisms of the cnidarian group (jellyfish and corals), whose morphology resembles those of a palm branch with a fleshy base acting as support on the seafloor (as shown in Fig. 3.8, right). Finally, the fossils with tri- or pentaradiate symmetry[43] (*Tribrachidium* and *Arkarua adami*, respectively) have been considered as belonging to the phylum of echinoderms (starfish, crinoids, sea urchins).

However, after this first explanation, the general view of the Ediacaria organisms has changed over the years. Common features of these taxa include their flattened morphology and, above all, the absence of common organs such as eyes or a mouth. The lack of a mouth is particularly controversial; it is unclear how a supposed trilobite or echinoderm precursor could have ingested food. Even fossils considered precursors of the simplest forms, such

[43] An organism (or object) has a triradial symmetry when it is possible to "cut" it into three equal parts, like a round cake. To have a pentaradial symmetry, it must be able to be cut into five equal parts.

as jellyfish, present interpretational challenges. Fossils thought to represent jellyfish umbrellas do not withstand rigorous scrutiny: all these remains had the concavity facing upwards, whereas a stranded jellyfish would have the opposite orientation. The fact that all these circular structures contain sand suggests that they were the base of sessile organisms, with the sediment acting as a stabilizer to prevent being transported or overturned by currents. As a result, most of the fossils once identified as jellyfish have been reclassified.

What kind of organism could have had a sand-filled base? Another common peculiarity of Ediacaran organisms is their flattened shape, providing a large surface area. Combined with the absence of a mouth, this feature has led specialists to propose various hypotheses. For example, researcher Adolf Seilacher, in the 1980s (Seilacher 1984), provided an original interpretation of these Precambrian communities. According to him, these creatures did not belong to any modern phylum and may not have been animals in the true sense. The anatomical trait that characterizes these organisms (and explains their particular way of life) is their flattened configuration. They all resemble pneumatic "mats" lying on the sediment (or partially buried) or erected on a stem, like *Charniodiscus*. This peculiarity, added to the lack of a functional mouth, does not correspond to any modern anatomy. Seilacher proposed that these organisms formed a distinct group of Ediacaran life (with some younger taxa persisting at least until the Middle Cambrian) but which subsequently became extinct. He named these creatures "vendobionts".

How could such organisms survive without mouths or eyes, simply lying on sediments or standing like flags? Seilacher suggested that they were photosynthesizers or lived in symbiosis with algae or other photosynthetic organisms. Those living in deeper, low-light environments could have been chemotrophs or osmotrophs.[44]

Consequently, these supposed animal precursors of Palaeozoic organisms (trilobites, echinoderms, molluscs) are now considered enigmatic creatures from a separate, completely extinct branch. Other researchers believed that they were lichens or colonies of single-celled or even giant protozoa, like today's Xenophyophorea.[45]

---

[44] Let me remind you that chemoautotrophic organisms are those capable of synthesizing their own food (glucose) through appropriate chemical processes (for example, sulphate-reducing microbes). Osmotrophic organisms feed on dissolved substances, which enter their cells through osmosis (that is, by simple diffusion of small molecules across the cell membrane. Neither of these dietary strategies requires a functional mouth.

[45] These organisms are giant protists that inhabit seas and oceans at great depths (below 500 m). Their growth patterns (a period of growth followed by a more or less long stasis interval before the cycle resumes) would have been identified in certain Ediacarian fossils, in particular, certain specimens of *Ediacaria* and *Mawsonites*. For this reason, some researchers belive that Xenophyophorea

Mark A.S. McMenamin, who coined the term "Ediacara's garden" to describe the biological communities of this period (highlighting the supposedly predator-free, idyllic conditions) proposed a personal hypothesis about the nature of these creatures (McMenamin 1998), in the late 1990s. According to him, they were multicellular organisms that, although sharing a common ancestor with modern animals, belonged to a completely different branch that became extinct after 500 My. Their unique anatomy may reflect an embryonic development entirely distinct from that of true animals. In addition, the Ediacaran creatures, completely lacking an oral opening, likely relied on photosynthesis, chemosynthesis, or osmosis for metabolism. According to McMenamin, these organisms constitute a unique group that is not fully animal in the modern sense. So why not call them "vendobionts", as Seilacher did? Ediacaran faunas thus represent an evolutionary experiment in multicellularity that flourished during this period but was unable to compete with emerging Palaeozoic organisms or became extinct due to still unknown causes.

Of course, not all Ediacaran organisms were vendobionts. Alongside the ubiquitous bacteria and stromatolitic structures, there were protists, including, most likely, "giant" forms such as the Xenophyophorea mentioned earlier. Naturally, the majority of macrofossils belong to vendobionts, which apparently constituted the dominant forms. However, alongside them are traces of true animals. Their presence is attested by the remains of sponges as well as by traces of biological activity in the sediments, possibly produced by organisms similar to worms or arthropods.[46]

Before leaving the Ediacarian and its extraordinary organisms, I would like to present one last hypothesis regarding the lifestyle of these communities. Recent analyses suggest that vendobionts may have been terrestrial rather than aquatic, living in a manner similar to lichens. Of course, not everyone agrees with this proposal,[47] but it has the advantage of encouraging researchers to test their reconstructions increasingly rigorously, using multiple methods simultaneously.

were genuinely present in the fauna of the time, and that a certain number of specimens must be attributed to them. See also Sect. "The Largest Eukaryotic Cells", in Appendix.

[46] However, it should be noted that the traces of animal activity found in Proterozoic strata are relatively simple. In general, they consist of horizontal tubes and rarely vertical ones (Erwin and Valentine 2013). This is far from the complexity of the trace types found in sediments starting from the Cambrian!

[47] The hypothesis that vendobionts (or at least some of them) may have been terrestrial organisms was put forward by Retallack (2013). His proposal found supporters (Knauth 2013) but also detractors (Xiao 2013).

Nonetheless, a question remains open: how did these organisms protect themselves from cosmic radiation in the absence of water? Today, we are shielded by the ozone layer (O_3), formed due to the abundance of gaseous oxygen (O_2) in our atmosphere. Although the oxygen content had already increased,[48] was it sufficient to allow vendobionts to emerge from the water? Or did they have other strategies to cope with the problem? Only time (and further research) will allow us to fully appreciate these ancient representatives of a vanished biodiversity and their ways of life on Earth at the end of the Proterozoic.

3.6 Summary of This Chapter

Towards the end of the Archean, the presence of eukaryotes in the biological communities of the time is evidenced by multiple chemical indicators. According to the theory most widely accepted by biologists and evolutionary scientists (developed by Lynn Margulis) the emergence of this new biological entity resulted from a symbiotic association of several prokaryotes, including at least one archaeon (which would have provided the "body" of the eukaryotic cell) and some bacteria (which later transformed into mitochondria and chloroplasts). This represents one of the earliest examples of interactions between different living organisms. However, not everyone is fully convinced by the symbiotic theory, and recent research suggests possible modifications, such as a viral origin of the eukaryotic nucleus, or proposes alternative hypotheses.

The eukaryotic cell differs from the prokaryotic cell in the presence of a true nucleus and, above all, in having organelles, intracellular structures enclosed by membranes and containing their own genetic material. This anatomical differentiation results in a fundamental functional difference. In prokaryotes, metabolic processes are centralized within the cytoplasm, whereas in eukaryotes, some essential metabolic transformations, such as energy production, occur in the mitochondria, thereby increasing efficiency. The energetic surplus generated can then be allocated to growth, division, and other cellular activities. This allows eukaryotic cells to acquire capabilities absent in prokaryotes, including the formation of multicellular organisms and the development of sexual reproduction.

---

[48] Chemical analyses indicate that during the last stage of the Proterozoic, oxygen levels could have increased to 5–18% of present value (Canfield and Teske 1996; Holland 2006). This study also suggested the existence, on the seabed at that time, of prokaryotic communities capable of sustaining themselves through the reduction of oxidized sulphates. For these processes to occur, oxygen levels must have reached an adequate value (see footnote 35 of this chapter).

The emergence of multicellularity enables eukaryotes to occupy ecological niches inaccessible to prokaryotes, largely due to their increased size and the development of organs and tissues specialized for different functions. Sexual reproduction, on the other hand, fundamentally changes the way descendants are produced. Through the interaction of two individuals of the same species, offspring are generated with unique genetic compositions: derived from their parents but distinct from both. Sexuality allows unprecedented genetic reshuffling within populations, thereby accelerating evolution. The earliest known fossils of multicellular organisms capable of sexual reproduction date to approximately 1 billion years ago.

Despite these achievements, eukaryotes are by no means superior to prokaryotes, neither today (considering species diversity or ecological dominance) nor likely in the past. Eukaryotes occupy ecological niches different from those of prokaryotes, reducing direct competition.

Both prokaryotes and eukaryotes successfully endured the transformations and climatic changes of the Precambrian. Eukaryotes, bacteria, and archaea survived glaciations, the assembly of continents into Rodinia and its subsequent breakup, and the first major global environmental change: the accumulation of oxygen resulting from the metabolism of photosynthetic organisms. Initially, oxygen accumulated in the oceans and later in the atmosphere, gradually shifting its composition toward modern levels, which would not be fully achieved until the Palaeozoic.

By the end of the Proterozoic, following glaciations events and increased oxygen levels, the first animals, fungi, and precursors of land plants began to inhabit the planet. Various groups of unicellular eukaryotes, the protists, also emerged. Bacteria and archaea continued their evolution alongside these eukaryotic organisms, often less conspicuously but persistently.

Appendix

The Largest Eukaryotic Cells

Excluding eggs, what are the largest single-celled eukaryotes existing today? Undoubtedly, *Gromia sphaerica* ranks among the largest. This spherical amoeba can reach a diameter of up to three centimetres. Imagine a single-celled organism clearly visible to the naked eye! It was discovered in the Arabian Sea, buried in the mud at a depth of more than 1 000 m (see the entry "Gromia sphaerica" of the digital encyclopaedia Wikipedia). This amoeba moves across the seafloor, leaving behind characteristic traces. Similar footprints have been identified in sediments dating from 2 to 1.8 billion years ago (Bengtson and Rasmussen 2009), suggesting that similar organisms existed at that time.

Even larger are the Xenophyophorea, giant unicellular eukaryotes (*Syringammina fragilissima*, for example, can reach a diameter of 20 cm!). These organisms inhabit the deep sea and possess an external skeleton composed of agglutinated sediment grains (see the entry "Xenophyophorea" of the digital encyclopaedia Wikipedia).

Fossil remains of these organisms are not yet confirmed. However, structures produced by Xenophyophores closely resemble certain Cambrian fossil traces, such as *Palaeodictyon*, as well as other traces found in more recent sediments (Levin 1994). In paragraph 3.5, I will revisit these organisms in relation to interpretations of some Ediacaran fossils.

Bacteria that Associate

Among bacteria, particularly cyanobacteria, filamentous arrangements are well known, in which all the cells are aligned in single file, in direct contact with one another. However, this arrangement is purely spatial and does not reach the level of organization seen in eukaryotic cell aggregations. In prokaryotes, such patterns mainly reflect the consequences of cell division.

In some bacteria, spherical assemblies can form, reaching diameters of a few tens of μm. Within these structures, some researchers have suggested the presence of metabolic differentiation among cells occupying different regions of the assembly. In a sense, this represents something more complex and structured than simple bacterial filaments. Nevertheless, these structures remain far from eukaryotic organization, in which different cells show clear coordination and specialized roles. Interestingly, spherical structures resembling these bacterial assemblies have been identified in geological layers dating from 2.5 to 1.6 billion years ago (Castanier et al. 1994).

Another notable example is the behaviour of myxobacteria, which is even more elaborate. Similar to myxomycetes, these prokaryotes form multicellular aggregates under nutrient stress, producing spores to disperse the next generation. Cohesion between cells appears to be maintained via real chemical signalling. Can we speak of embryonic multicellular behaviour among bacteria? It is still too early to say definitively. However, ongoing research in this field may help clarify the situation. Moreover, some microbiological studies are beginning to highlight the "social behaviours" exhibited by prokaryotes (Bapteste 2013). It is therefore possible that the world of prokaryotes may still hold some fascinating surprises in this regard.

Why Does Cell Division Exist in Bacteria?

The Italian physicist Mario Ageno attempted to answer this question by proposing a plausible hypothesis. His reasoning stemmed from studies of the growth of isolated bacterial cells. Ageno's experiments demonstrated that this growth is not linear, as is often suggested in the biological literature, but exponential (Ageno 1989; Ageno et al. 1990). In particular, the bacterial surface, through which nutrients enter via specialized "channels," grows exponentially. This has profound implications for cellular functioning because the amount of substances entering the bacterium also increases exponentially, being proportional to the surface area. However, the molecules responsible for metabolizing these substances (enzymes) only increase linearly, produced from a single strand of the genome, at best doubling after DNA replication. Since an exponential function eventually outpaces any linear (or fixed-exponent) function, nutrient intake would surpass the cell's metabolic capacity once enzyme synthesis reaches its maximum. This imbalance could compromise cellular integrity, potentially threatening survival.

Ageno proposed a practical solution: the cell divides in two before excess materials accumulate. Division produces two daughter cells whose size (surface area) is compatible with the processing capacity of the available enzymes. Growth then resumes, again following an exponential law for nutrient intake and a linear law (doubling after DNA replication) for enzyme production. This cycle repeats with each division to prevent the accumulation of unprocessed substances.

This hypothesis was presented in Ageno's final book (Ageno 1992a, b). The author himself acknowledges that the idea still lacks experimental verification. Nevertheless, it is worth emphasizing that Ageno's rigorous methodology in cell biology allowed him to formulate original propositions that could significantly advance the field.

References[49]

Ageno, M. 1989. La crescita batterica III. La legge di crescita del singolo batterio. *Atti Accademia Dei Lincei—Rendiconti Scienze Fisiche e Naturali* 83: 335–341.

Ageno, M. 1992a. *La macchina batterica*. Rome: Lombardo editore.

Ageno, M. 1992b. *Punti cardinali*. Milan: Sperling & Kupfer.

[49] Works preceded by an asterisk can be read, more or less easily, by people with a non-professional skilling.

Ageno, M., A. Battistini, E. Liberati, and A.M.F. Valli. 1990. Verifiche sperimentali della teoria della crescita batterica I. *Atti Accademia Dei Lincei—Rendiconti Scienze Fisiche e Naturali* 1: 55–62.

*Archibald, J. 2014. *One plus one equals one—Symbiosis and the evolution of complex life*. Oxford University Press.

Ashkenazy, Y., H. Gildor, M. Losch, F.A. Macdonald, D.P. Schrag, and E. Tziperman. 2013. Dynamics of a snowball earth ocean. *Nature* 495: 90–93.

*Bapteste, E. 2013. *Les gènes voyageurs—L'odyssée de l'évolution*. Paris: Belin.

*Barbault, R. 2006. *Un éléphant dans un jeu de quills—L'homme dans la biodiversité Seuil*. Paris: Points Sciences.

Baum, D.A., and B. Baum. 2014. An inside-out origin for the eukaryotic cell. *BMC Biology* 12: 1–22.

Bengtson, S., and G. Budd. 2004. Comment on small Bilaterian fossil from 40 to 55 million years before the Cambrian. *Science* 306: 1291.

Bengtson, S., and B. Rasmussen. 2009. New and ancient trace makers. *Science* 323: 346–347.

Brocks, J.J., G.A. Logan, R. Buick, and R.E. Summons. 1999. Archean molecular fossils and the early rise of eukaryotes. *Science* 285: 1033–1036.

Butterfield, N.J. 2000. *Bangiomorpha pubescens* n. gen., n. sp.: Implications for the evolution of sex multicellularity, and the Mesoproterozoic/Neoproterozoic radiation of eucaryotes. *Paleobiology* 26 (3): 387–404.

Canfield, D.E., and A. Teske. 1996. Late Proterozoic rise in atmospheric oxygen concentration inferred from phylogenetic and sulphur-isotope studies. *Nature* 382: 127–132.

Castanier, S., J.-P. Perthuisot, A. Maurin, V. Gèze, and G. Camoin. 1994. Colonies bactériennes organisées actuelles Quelques réflexions sur l'évolution des procaryotes. *Géobios* 27 (6): 645–657.

Chen, J.-Y., D.J. Bottjer, P. Oliveri, S.Q. Dornbos, F. Gao, S. Ruffins, H. Chi, C.-W. Li, and E.H. Davison. 2004. Small bilaterian fossil from 40 to 55 million years before the Cambrian. *Science* 305: 218–222.

*Collective Work. 2013. *Evolution—Les quatre premiers milliards d'années: De la cellule primitive à l'apparition des mammifères*. Geo savoir, hors-série 5 (February–March).

de Reviers, B. 2018. Les associations dans l'évolution du vivant. In *Microbiodiversité. Un nouveau regard*, ed. L. Palka, 51–103. Paris: Matériologique.

Degli Esposti, M., B. Chouaia, F. Comandatore, E. Crotti, D. Sassera, P.M.-J. Lievens, D. Daffonchio, and C. Bandi. 2014. Evolution of mitochondria reconstructed from the energy metabolism of living bacteria. *PLoS ONE* 9 (5): 1–22.

El Albani, A., S. Bengtson, D.E. Canfield, A. Bekker, R. Macchiarelli, A. Mazurier, E.U. Hammarlund, P. Boulvais, J.-J. Dupuy, C. Fontaine, F.T. Fürsich, F. Gauthier-Lafaye, P. Janvier, E. Javaux, F. Ossa Ossa, A.C. Pierson-Wickmann, A. Riboulleau, P. Sardini, D. Vachard, M. Whitehouse, and A. Meunier. 2010.

Large colonial organisms with coordinated growth in oxygenated environments 2.1 Gyr ago. *Nature* 466: 100–104.

Erwin, D.H., and J.W. Valentine. 2013. *The Cambrian explosion—The construction of animal biodiversity*. Greenwood Village (Colorado): Roberts and Company.

Fedonkin, M.A. 2003. The origin of the metazoa in the light of the proterozoic fossil record. *Paleontological Research* 7 (1): 9–41.

Forterre, P. 1996. À la Recherche de LUCA. In *Meeting fondation de Treilles*.

*Forterre, P. 2007. *Microbes de l'enfer*. Paris: Belin pour la Science.

Forterre, P., and M. Gaïa. 2018. La place des virus dans le monde vivant: Le concept de virocell. In *Microbiodiversité. Un nouveau regard*, ed. L. Palka, 23–49. Paris: Matériologique.

Forterre, P., S. Gribaldo, and C. Brochier. 2005. Luca: À la recherché du plus proche ancêtre commun universel. *Médicine Sciences* 21 (10): 860–865.

Génermont. J. 2014. *Une histoire de la sexualité. Plus d'un milliard d'années d'évolution*. Matériologique, Paris, 23–49.

Gibson, T.H., P.M. Shih, V.M. Cumming, W.W. Fischer, P.W. Crockford, M.S. Hodgskiss, S. Wörndle, R.A. Creaser, R.H. Rainbird, T.M. Skulski, and G.P. Halverson. 2018. Precise age of *Bangiomorpha pubescens* dates the origin of eukaryotic photosynthesis. *Geology* 46: 135–138.

Hagadorn, J.W., S. Xiao, P.C.J. Donoghue, S. Bengstron, N.J. Gostling, M. Pawlowska, E.C. Raff, R.A. Raff, F.R. Turner, Y. Chongyu, C. Zhou, X. Yuan, M.B. McFeely, M. Stampanoni, and K.H. Nealson. 2006. Cellular and subcellular structure of neoproterozoic animal embryos. *Science* 314: 291–294.

Han, T.M., and B. Runnegar. 1992. Megascopic eukaryotic algae from the 2.1-billion-year-old negaunee iro-formation, Michigan. *Science* 257: 232–235.

Hoffman, P.F., A.J. Kaufman, G.P. Halverson, and D.P. Schrag. 1998. A Neoproterozoic snowball earth. *Science* 281: 1342–1346.

Holland, H.D. 2006. The oxygenation of the atmosphere and oceans. *Philosophical Transactions of the Royal Society of London* 361: 903–915.

Horodyski, R.J., and L.P. Knauth. 1994. Life on land in Precambrian. *Science* 263: 494–498.

Huntley, J.W., S. Xiao, and M. Kowalewski. 2006. 1.3 Billion years of acritarch history: An empirical morphospace approach. *Precambrian Research* 144: 52–68.

Javaux, E.J., H.A. Knoll, and M.R. Walter. 2001. Morphological and ecological complexity in early eukaryotic ecosystems. *Nature* 412: 66–69.

Knauth, L.P. 2013. Not all at sea. *Nature* 493: 29.

*Knoll, H.A. 2004. *Life in a young planet—The first three billion years of evolution on the earth*. Princeton University Press.

Knoll, H.A., E.J. Javaux, D. Hewitt, and P. Cohen. 2006. Eukaryotic organism in proterozoic oceans. *Philosophical Transactions of the Royal Society of London* 361: 1023–1038.

*Lane, N. 2010. *Life ascending—The ten great inventions of evolution*. London: Profile Books.

*Langueney, A. 1987. *Le sexe et l'innovation*. Paris: Seuil.

Lecointre, G., ed. 2009. *Guide critique de l'évolution.* Paris: Belin.

Levin, I.A. 1994. Paleocology and ecology of xenophyophores. *Palaios* 9 (1): 32–41.

Li, C.-W., J.-Y. Chen, and T.-E. Hua. 1998. Precambrian sponges with cellular structures. *Science* 279: 879–882.

Loron, C.C., François, C., Rainbird, R.H., Turner, E.C., Borensztajn, S., and Javaux. E.J. 2019. Early fungi from the Proterozoic era in Arctic Canada. *Nature* 570: 232–235.

Margulis, L. 1999. *Symbiotic planet—A new look at evolution.* New York: Basic Books.

*McMenamin, M.A.S. 1998. *The garden of Ediacara—Discovering the first complex life.* New York: Columbia University Press.

*Mieli, E., A.M.F. Valli, and C. Maccone. 2025. *The living galaxy—Winners and losers in the milky way.* Cham, Switzerland: Springer Nature.

Ninio, J. 1996. Gene conversion as a focusing mechanism for correlated mutations: A hypothesis. *Molecular and General Genetics* 251: 503–508.

Pennisi, E. 2013. The man who bottled evolution. *Science* 342: 790–793.

Porter, S.M. 2004. The fossil record of early eukaryotic diversification. *Paleontological Society Paper* 10: 35–50.

Porter, S.M. 2006. The proterozoic fossil record of heterotrophic eukaryotes. In *Neoproterozoic geobiology and paleobiology*, ed. S. Xiao and A.J. Kaufman, 1–21. Berlin: Springer.

Retallack, G.J. 2013. Ediacaran life on land. *Nature* 493: 89–92.

Rogers, J.J.W. 1996. A history of continents in the past three billion years. *The Journal Geology* 104: 91–107.

*Sansjofre, P., and G. Hir. 2003 Le paradoxe de la Terre boule de neige. *Pour la Science* 486: 26–35.

Seilacher, A. 1984. Late Precambrian Metazoa: Preservational or real extinctions? In *Patterns of change in earth evolution*, ed. D.H. Holland and A.F. Trendall, 159–168. Berlin: Springer.

*Selosse, M.A. 2017. *Jamais seul—Ces microbes qui construisent les plantes, les animaux et les civilisations.* Arles: Actes Sud.

Strother, P.K., L. Battiston, M.D. Brasier, and C.H. Wellman. 2011. Earth's earliest non-marine eukaryotes. *Nature* 473: 505–509.

Taylor, T.N., E.L. Taylor, and M. Krings. 2009. *Paleobotany: The biology and evolution of fossil plants*, 2a ed. Amsterdam: Academic Press.

Voigt, O., A.G. Collins, B.V. Pearse, J.S. Pearse, A. Ender, H. Hadrys, and B. Schierwater. 2004. Placozoa—No longer a phylum of one. *Current Biology* 14 (22): 944–945.

Watson, J.D., N.H. Hopkins, J.W. Roberts, J. Argetsinger Steitz, and A.M. Weiner. 1987. *Molecular biology of the gene*, vol. 2, 4th ed. San Francisco: The Benjamin/Cummings Publishing Company.

Xiao, S. 2013. Muddying the waters. *Nature* 493: 28–29.

Xiao, S., Y. Zhang, and A.H. Knoll. 1998. Three-dimensional preservation of algae and animal embryos in a neoproterozoic phosphorite. *Nature* 391: 553–558.

Yuan, X., S. Xiao, and T.N. Taylor. 2005. Lichen-like symbiosis 600 million years ago. *Science* 308: 1017–1020.

4

The Palaeozoic, the Era of "Ancient Life"

4.1 The Explosion of Life at the Cambrian

With the rise of multicellular organisms and the revolution brought about by sexual reproduction, some of the most important biological milestones in the eukaryotic domain were achieved. Of course, evolution did not stop there: new animals would replace the old, and the same occurred in plants, fungi, and microorganisms, with new arrivals ready to conquer emerging environments… However, these are changes in the actors, but the roles remained largely the same.

Why, then, continue this story up to the present day? What new and significant innovations are left to describe? In reality, the history of biodiversity cannot be halted at this stage, for two main reasons:

(1) The scope of biodiversity. As commonly understood, biodiversity includes all living beings that people can easily see and appreciate, especially plants and animals. Important fossil traces of such organisms first appear in the Phanerozoic, the time interval that begins where we left off in the previous chapter and extends to the present. Stopping here would deprive readers of the means to connect the earliest stages of life's history with the more recent ones.

(2) The rise of animal societies. To fully understand the evolution of living beings, it is necessary to explore another crucial milestone: the emergence of animal societies. These mark a new level of interaction among living organisms. Although societies may have existed before the Phanerozoic,

A. M. F. Valli, *The Three Domains of Life*, Copernicus Books, https://doi.org/10.1007/978-3-032-14802-5_4

no evidence of them is found in Precambrian records. Instead, they begin to appear in later epochs, which will be covered in the following chapters.

In this and the next two chapters, I will describe the events of the Phanerozoic. This vast interval is divided into three eras: the Palaeozoic (the focus of this chapter), the Mesozoic, and the Cenozoic.[1]

The Palaeozoic (also called the "Primary Era" or simply "Primary") began around 541 million years ago. Its onset is marked by major changes in flora and fauna, especially the appearance of new organisms, identifiable through the fossils they left behind.[2]

The most spectacular phenomenon of the Phanerozoic strata is undoubtedly the abundance, variety, and complexity of fossils compared to those of earlier eras. This is the so-called "Cambrian Explosion" or "Cambrian Life explosion",[3] a term describing the sudden (though in reality it may have taken 20–30 million years) appearance of a large number of animal groups, both extinct and still living today, at the beginning of the Cambrian. In fact, while many of the strange creatures of the Ediacaran disappeared at the start of the Primary, many more appeared, especially the "bilaterians". What kind of animals are they? As the name suggests, they are all those animals with bilateral symmetry, whose right side mirrors the left. This group includes nearly all fossil and modern animals of the Phanerozoic, with the exception of sponges, coelenterates (jellyfish, hydroids, sea anemones, and corals), and ctenophores.[4] The latter are marine animals resembling jellyfish. Once classified as coelenterates, they differ in lacking stinging cells (cnidoblasts), using instead adhesive cells to capture prey. Their earliest known fossils date back to the early Cambrian.[5]

[1] The limits of my book do not allow me to present, in a comprehensive way, the total evolution of biodiversity during the Phanerozoic. However, the reader interested in having a somewhat more complete overview (especially in terms of animal evolution), but still simple, reading Gould et al. (1993).

[2] For the specialists in this field, the Precambrian/Cambrian boundary is marked by the appearance of very precise traces of activity (characteristics, for example, because they are vertically arranged through the sediments) unknown in the earliest periods (Sour-Tovar et al. 2007).

[3] The Cambrian is the first period of the Palaeozoic. It is between 541 and 485 My. The Cambrian is followed, in order, by the Ordovician (485–444 My), the Silurian (444–419 My), the Devonian (419–359 My), the Carboniferous (359–299 My) and the Permian (299–252 My), with which it came to its end.

[4] Placozoa are considered as the simplest animals; they are also excluded from the bilaterians. These living being have already been introduced in footnote 40 of Chap. 3.

[5] The fossil remains of the oldest ctenophores date from the Lower Cambrian and come from China (Chen et al. 2007; Hou et al. 2008).

Let us go back to the bilaterians. Must groups have appeared during the Cambrian (between 541 and 485 My). In particular, all major phyla[6] of animals with mineralized tissues (skeletal elements such as shells, plates, bones) appear in the first half of this period. All or almost all of them.[7] Because mineralized tissues resist decay, these organisms fossilized more readily than soft-bodied forms, making them disproportionately well represented in the fossil record.[8]

This clarification is important and it raises an important question: was the Cambrian Explosion a true evolutionary event, or merely an artefact of fossilization? In other words, did biodiversity genuinely increase dramatically, or do we simply detect it more easily because mineralized animals[9] were more likely to be preserved? In the latter case, it would not be correct to speak of an explosion of Life in general. In fact, this phenomenon could be explained by the fact that organisms capable of producing biomineralizations[10] (and, therefore, susceptible to fossilization more easily) become more abundant. All this, without the need for total biodiversity to increase!

Of the two possibilities, which is the most relevant? Experts remain divided (Levinton 2008), but what is certain is that organisms secreting rigid tissues became much more common at this time! But why? What happened from this moment? What has changed on the planet?

Various explanations have been proposed. One influential idea, put forward by Andrew Parker (1998), highlights the role of light and vision: the rise of predators with functional eyes would have pressured prey to develop new defences like spines, shells, and protective armour.

[6] The higher taxonomic groups are known as "phylum" (plural "phylia"). They refer to groups of animals that have the same general anatomical plan. For example, the possession of a chitinous coating and articulated legs defines arthropods; the possession of the mantle (part of the body capable of secreting a calcareous shell) and the radula (a kind of rough tongue, endowed with chitinous growths—attention, the radula can be lost during the course of evolution) characterizes molluscs.

[7] The bryozoa, small polyp-shaped animals living in colonies covered with an external skeleton, are missing. The oldest known fossil evidence of such animals dates from the Ordovician (Valentine et al. 1999). However, it must not be forgotten that several modern forms are devoid of mineralized tissues and that the oldest fossils already have an anatomical organization which suggests a previous evolution.

[8] Nevertheless, many fine soft tissue fossils typical of groups comprising animals with no mineralized parts have been found in exceptional deposits where even the most delicate structures can be preserved. Limiting myself to the Cambrian sites, I remember those of Chengjiang in south-west China (Hou et al. 2008) and that of Burgess in Canada (Briggs et al. 1994).

[9] It should be remembered that some Ediacarian organisms, which we have seen in paragraph 3.5, already had rigid tissues, although their degree of mineralization was lower than Palaeozoic ones.

[10] Biomineralization is the ability of living things to secrete rigid tissues. An example of biomineralization is the shell of a snail, secreted from the animal's coat.

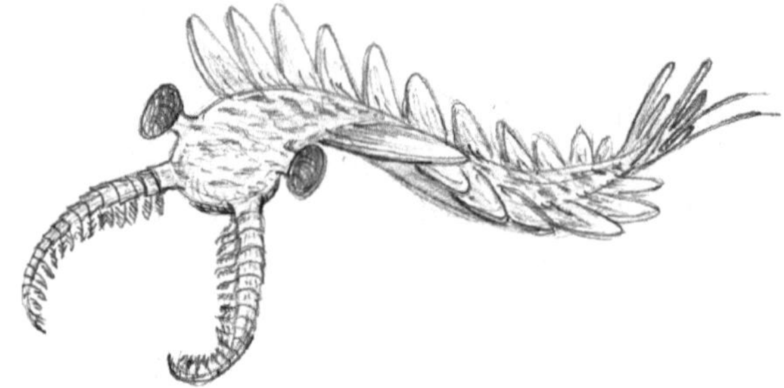

Fig. 4.1 Reconstruction of *Anomalocaris*, one of the first large predators of the prehistoric seas. The species depicted could reach, on average, half a metre in length

Fossil evidence supports this, with the appearance of well-developed eyes[11] in Cambrian arthropods as well as clear signs of predation on shells. In addition, we finally find the remains of early predators: their vestiges have been preserved in the sediments to the present day. Among them, different from all modern creatures, I remember *Anomalocaris* and his kind. *Anomalocaris* (Fig. 4.1) was a shrimp-like creature with prominent eyes and grasping appendages for grabbing prey. It moved in the water by swimming thanks to undulations of its body, also helped by the lateral lobes that could function as fins.

Anomalocarids were not only the oldest predatory organisms for which fossils are available, but they were also the largest creatures of their time: they measured between 50 cm and 2 m in total length, real giants compared to other animal! They were undoubtedly the most feared marine predators of their time and are known in the Lower and Middle Cambrian of various locations in Asia, North America, Europe and even Australia. The youngest representative of the group was exhumed in Moroccan sediments dated between 488 and 472 My (Lower Ordovician; Van Roy and Briggs 2011). Later, they were replaced by other types of predators: cephalopods, sea scorpions, and fish.

[11] The evolutionary history of the eyes has not yet been fully elucidated. For example, did the eyes of vertebrates have a common origin with those of annelids or molluscs? The possession of a common gene complex among all animals with functional eyes would tend to favour the former, although this is not yet settled. In fact, each major group has eyes that are structurally different from one another. The first stage of evolution of the eye would be the formation of a photoreceptor cell sheet (capable of detecting light) wedged between one layer of pigmented cells and another, transparent cells, which serves as protection. This structure, initially flattened, would later curve to become globular. The evolution of the eye may then have taken different paths in each phylum. Among the simplest eyes, I cite those of planarians (flat worms). The oldest known fossil eyes were those of an arthropod. For a brief history of the eyes, see the Wikipedia digital encyclopaedia, at the entry "Evolution of the eye".

As indicated, this animal possessed perfectly developed and functional eyes. However, the eyes are not the only organs that allow predators to see prey. Chemical or mechanical signals also allow prey to be found, even over distances greater than those allowed by the eyes, especially in the aquatic environment. And the response of potential victims is not necessarily to "armour" themselves with highly mineralized tissues. Chemical defences can be just as effective, if not better, even for those creatures that cannot leave the substrate on which they live. Parker's hypothesis is undoubtedly seductive but, in my opinion, it is not enough to completely solve the problem. The eyes certainly had their importance. However, as in most cases, it is possible that several causes have been added to promote the secretion of rigid tissues.

In fact, recent studies related to the Cambrian explosion have provided a list of a series of events and/or evolutionary achievements that, acting in unison, would have laid the foundations for Cambrian biodiversity (Marshall 2006; Levinton 2008; Erwin and Valentine 2013).

First, a change in the "chemistry of the oceans". This does not mean that the chemical reactions of the Precambrian followed different laws from the current ones, nor that the composition of the waters was radically different from today. Simply, the waters of the oceans could be enriched by chemical elements, transported by rivers from the continents or acquired in a different way. It seems likely, for example, that calcium carbonate precipitation became increasingly easy, between the Ediacarian and the early Cambrian. Small variations in the content of some element, and also in the temperature or pressure of the water, can, in the long run, affect the solubility of limestone and, therefore, the ability of marine creatures to produce biomineralizations. See the explanation dedicated to Section "Variation of Biomineralization in the Marine Environment", in Appendix.

From this point of view, the reduction of the carbon dioxide (CO_2) content is essential, because this gas helps to increase the acidity of the water and to counteract the formation of calcium carbonate, one of the main elements of biomineralization. Oxygen, on the other hand, plays a role opposite to that of CO_2. The content of this element has not ceased to increase in different environments, as a consequence of the photosynthesis of the first cyanobacteria. At more than 600 My by the present time, deep marine environments had already begun to spread, below the limit of light penetration in the water (Canfield et al. 2007). This increase would have allowed the first animals to be able to colonize such depths. In fact, we have to recognise that the oxygen content has been crucial for the spread and diversification of the group. This gas was not only vital for respiration but also for the synthesis of collagen, a key tissue in metazoans (= animals; Canfield and

Teske 1996). Collagen, in order to be synthesized, needs a certain concentration of oxygen. Below this limit, the reaction does not occur. Although an increase in oxygen also has deleterious secondary effects (for example, the inhibition of various elements and chemical molecules crucial for metabolic processes), the importance of this gas for the evolution of animals needs no further confirmation. Rather than being a "permissive factor", the increase in oxygen would have acted as a force that would direct evolution towards complex biological systems (Fedonkin 2003). In any case, it was an essential phenomenon to counteract the action of CO_2 and to promote the production of biomineralization based on calcium carbonates.

Another important event to explain the phenomena that occurred at the beginning of the Palaeozoic was, without a doubt, the development of the main organs of animals, such as the eyes (but, probably, not only them). Let us remember that the creatures that abounded during the previous era, after detailed analysis, seemed to lack not only eyes, but also mouths and many other sensory organs (hence the hypothesis regarding their particular phylogenetic state presented in Sect. 3.5). The first creatures equipped with all these organs are encountered during the Cambrian (but this does not mean that there were no other, even more ancient ones!).

Finally, it is also necessary to take into account the evolution of genes or particular complexes of genes capable of modifying the development of the embryo. The presence of these genetic characteristics and their modifications are properties that are difficult to trace through the simple study of fossils. In all cases, highlighting them is much more complex than identifying particular organs in living beings or distinctive elements in environments. Only the analysis of modern organisms can help us reconstruct the evolution of entire gene complexes during the epochs of the past.

The combination of all these factors (and, of course, others as well) could have constituted the "triggering" element of the phenomena observed at the beginning of the Primary. The interplay of these factors produced an extraordinary diversification of life forms and ecological interactions, unprecedented compared to earlier times.[12]

An example of the spread of animal groups is given by trilobites, undoubtedly the organisms that more than others characterize the entire Palaeozoic.[13]

---

[12] To get a glimpse of the Cambrian creatures and appreciate their interactions, I recommend the beautiful books by Erwin and Valentine (2013) and Gould (2000).

[13] Here are some works on trilobites: Levi-Setti (1995), Fortey (2000), Lebrun (1995) and Bonino and Kier (2010). If you like to browse the web, you can see the entry "Trilobita" of the digital encyclopaedia Wikipedia.

These arthropods, exclusively fossil, whose name derives from the tripartition of their body,[14] appear during the Lower Cambrian, at the base of the Cambrian explosion (around 521 My), and appear among the first animals with eyes. Trilobites flourished throughout the Palaeozoic until their extinction at the end of the Permian, producing over 10,000 known species. Ranging from a few millimetres to over 70 cm, they filled diverse ecological niches and displayed remarkable morphological variety, much like modern insects and crustaceans (Fig. 4.2).

Though trilobites are emblematic, they represent only one facet of the rich biodiversity that emerged during the Palaeozoic, both in aquatic and terrestrial environments.

Before concluding this paragraph, I want to add some notes about certain fossils we have already encountered: the famous stromatolites. These structures have accompanied us since the observation of the first traces of life on Earth. Of course, they are also present during the Palaeozoic. In the Allier department (where I live and wrote these pages), especially near the city of Souvigny, it is sometimes possible to collect rocks that, in section, display the undulations typical of these bio constructions. These are Carboniferous–Permian freshwater stromatolites, formed when France was located near the Equator.

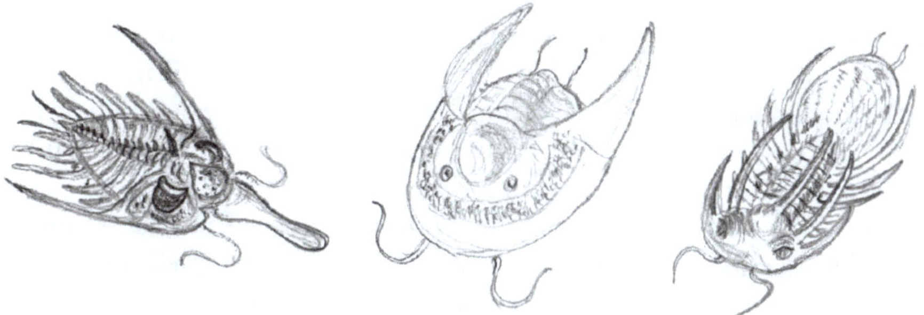

Fig. 4.2 Brief glimpse of the biodiversity of Moroccan trilobites dating from 411 and 393 My (Middle Devonian): on the left, *Psychopyge elegans*, a species with a very pronounced rostrum (about 10 cm in total length); in the middle, a representative of the genus *Harpes* (about 5 cm in total length); on the right, *Kolihapeltis rabatensis*, with long lateral spines and three spines behind the head (about 6 cm in total length)

[14] The name trilobite means "three-lobed". This designation refers to the longitudinal division of the animal (head, cephalon; chest, thorax; terminal part, pygidium) and to the longitudinal one, with two lateral expansions, one per side, around the central axis. For the definition of the word arthropod, see footnote 6 to this paragraph.

4.2 Past and Present Echinoderms

Due to the differentiation (real or perceived) of animals and the abundance of major groups throughout the Early Palaeozoic, seas and oceans teemed with life. Marine biodiversity reached a peak compared to the previous era. Thanks to ecological differentiation, Palaeozoic waters appeared different from Precambrian ones. Predators now roamed the waters in search of prey or ambushed their victims hidden on the seabed. Mobile creatures grazed on algae or bacterial mats. Others, more or less active, walked along the seabed searching for food particles hidden in the sand or mud. Some were less selective, swallowing everything to extract tiny food items. Finally, some organisms waited passively for food to rain down, anchored to the substrate at varying heights above the seafloor.

I could discuss trilobites, fossil animals unable to cross the Palaeozoic boundary, or priapulid worms,[15] curious marine animals still existing today, buried in the sediment. However, I have chosen to focus on echinoderms: sea urchins, starfish, and related types.

Aside from trilobites, this phylum is particularly suitable to represent marine diversity during the Palaeozoic. Today, there are five major echinoderm divisions[16] (classes) in today's seas and oceans, but during the Early Palaeozoic, there were 21 (Ubaghs 1967)!

Before discussing their past biodiversity, let us introduce the group and understand its defining characteristics. While stars and sea urchins are well-known, ophiuroids, sea cucumbers, and crinoids are less familiar.

Sea cucumbers, as their name implies, resemble cucumbers. Their body walls are soft, reinforced by small calcareous speckles. Their mouth is surrounded by branching tentacles for capturing food and bringing it to the mouth. They inhabit sandy or muddy bottoms, and some abyssal forms have learned to swim above the seafloor. Sea cucumbers are relatively abundant in tropical and deep waters. This brief glimpse gives us an idea of the modern diversity of sea cucumbers which, at least, in ecological terms, is far from negligible!

---

[15] Priapulids are marine worms characterized by the possession of a spiny proboscis extensible. All modern species are carnivorous: hidden on the seabed, they capture prey, mainly ringlets. This is a minor group, of some dozen current species. However, their fossil representatives are known since the Cambrian.

[16] The five current echinoderms include echinoids (Echinoidea; sea urchins), sea cucumbers (Holothuroidea; sea cucumbers), asteroids (Asteroidea; starfish), ophiuroids (Ophiuroidea; serpentine stars) and crinoids (Crinoidea; sea lilies). Some time ago, a sixth group was also recognized, that of the Concentricycloidea, including the genus *Xyloplax*, an enigmatic small creature of circular shape. However, recently, these animals have been recognised as belonging to the asteroid class.

Crinoids, even stranger, superficially resemble plants (hence the name "sea lilies"). Current forms have five arms, which can branch multiple times, forming crowns with up to 200 elements depending on the species. The shapes that are typical of deep water have a long peduncle to anchor themselves to the substrate. Deep-water forms have a long stalk to anchor themselves; coastal forms lack a stalk but have "roots" attached to the calyx. Crinoids can swim by waving their arms synchronously.

Echinoderms are therefore very differentiated both morphologically and ecologically. But what do they have in common, to classify them all together? What are the characteristics of the group?

All echinoderms have a subcutaneous skeleton (just below the skin), made of calcite plates, which can be developed differently (for example, in sea cucumbers, this structure is reduced to simple non-junctive spicules, while in urchins or crinoids it encloses the whole body). They also possess a water-filled system (the water vascular system), enabling locomotion, adhesion, prey capture, and food transport; it may also assist respiration. Water enters via pores on the madreporite, a plate present in all echinoderms. Pores allow communication between the internal device and the external environment. They therefore allow the regulation of the hydrostatic pressure of the animal.

Many extant (but also fossil) adult echinoderms exhibit a pentaradiate symmetry (a radiant symmetry of order five). That is, the animal can be divided into five equal "slices" between them. This symmetry, although widespread, is not universally present in all echinoderms. This peculiarity has often posed problems for specialists interested in understanding the phylogenetic relationships of the different groups of the phylum. Some authors state that pentaradiate symmetry should not be considered as a characteristic of the group,[17] because it can be acquired secondarily in certain evolutionary lines. Instead, the really essential traits are the calcareous skeleton and, above all, the aquifer system. All animals that possess one would be, by definition, echinoderms. So, to correctly identify a fossil form, just make sure of the presence of such an apparatus.

Despite the rigidity of the definition, we have been able to see that modern species have a very variable morphology. If we now add that of fossil forms, we are bound to encounter great surprises. But let us proceed in order.

[17] The theory linked to the EAT model (Extraxial-axial theory) allows to explain the morphology of different echinoderms, modern and fossil, and to explain the phylogenetic relations between them (David and Mooi 1999).

Let us start by situating the appearance of echinoderms in time. It is a really old group. Echinoderms first appeared[18] in the Early Cambrian (approximately between 521 and 509 My), during an epoch that does not yet have an "official" name on the international timescale (it is referred to as "Series 2"). By comparison, the oldest known trilobites are, more or less, the same age!

The edrioasterolds are part of these ancient echinoderms. They are an exclusively Palaeozoic group. These organisms were formed by a more or less rigid calyx,[19] placed on the bottom or attached to the substrate. The oral part, where the mouth is located, had five surfaces, more or less curved, depending on the species, which formed a structure similar to the body of a starfish (Fig. 4.3). The surfaces in question are called "ambulacral" and constitute the system that allows the transfer of food to the mouth, located at the junction point. These animals seem to be spineless, unlike sea urchin, or real arms, like those of crinoids. Despite their apparently simple morphology, they thrived from the Cambrian until the end of the Carboniferous (almost as long as the entire duration of the Palaeozoic), with at least forty species, differentiated by the structure of the theca and the shape and proportions of the ambulacral surfaces.

The rhombiferans (Rhombifera), appear a little later. Although the oldest fossils can be traced back to the Cambrian (although we are not yet sure of their attribution), these animals are best known during the Ordovician. Their name comes from the arrangement of their aquifer system pores along a rhomboidal perimeter. The body consisted of a globular or sack-shaped calyx, consisting of plates (Fig. 4.3). Some could have appendages, but all species were equipped with pinnules, a sort of tiny tentacles whose function was to capture food and make it reach the mouth, as happens today in sea lilies. The rhombiferans also had a peduncle to anchor themselves to the bottom, remaining erect above it, in order to take advantage of the sea currents. In short, they might look like crinoids, but the structure of the calyx is different, as are the arms, when present. Their representatives lived throughout the Ordovician and Silurian, disappearing during the Devonian. Having lasted at least 65 million years, the group certainly cannot be considered as marginal.

But if we had to award a prize to the group that has the most bizarre morphology, the victory would, without a doubt, go to the stylophores

---

[18] The attribution to echinoderms of *Arkarua adami*, an Australian Edacarian form, is highly disputed. In fact, the arguments used to approach this phylum are based more on its pentaradiate symmetry than on precise anatomical characters. On the other hand, no one has yet pointed out the presence of an aquifer system on *Arkarua* or other Edacarian fossils!

[19] The echinoderms' calyx is a structure in the form of a receptacle corresponding to their body, where the main organs are kept. It is protected by more or less imbricate calcareous plates. The rigidity of the whole depends on the degree of mineralization of the plates and their vessel.

Fig. 4.3 Reconstruction of three Palaeozoic echinoderms. On the left, the edrioasteroid *Cyathocystis plautinae* (Middle Ordovician) fixed on a rock (diameter, about half a centimetre); in the centre, the "rhombiferans" *Pleurocystites filitextus* (Middle Ordovician), anchored to a roughness of the bottom (height, about 10 cm); on the right, the stylophore (Stylophora) *Ceratocystites perneri* (Middle Cambrian) resting on the sandy bottom with its appendage stretched out against the direction of the current (total length, about 5 cm)

(Stylophora).[20] This class includes echinoderms with a flexible articulated appendage inserted in a flattened theca, with variable morphology depending on the species. Often, the calyx had an asymmetry, even very pronounced, between the right and left sides. The appendix, which, at first glance, seems to be the tail, was actually located in the animal's front. There was also the mouth, located at the base of the appendix, towards the graft with the main body, and with the opening directed downwards. The appendage, in the furthest part from the animal's body, was more flexible than a its base. Equipped with movements to the right and left, she swept the seabed in search of food. Which, thanks to a groove in the lower surface of the appendix, was transported to the mouth. In reality, the appendix of the stylophores is equivalent in all respects to the arm of a pedunculated echinoderm. Unlike the latter, however, the stylophore did not live upright above the substrate to filter the plankton. Instead, he was lying on the bottom (Fig. 4.3), with his "arm" stretched out against the dominant currents, like a strange armoured worm, equipped with a trunk. At least since the end of the Cambrian, these animals have been found in sediments considered typical of relatively deep and cold marine environments. In these confined

[20] For information on the Stylophora, about their diversity and ecology, see the text by Lefebvre (2003, 2022).

environments, they were able to limit competition from other depositional organisms[21] present in the shallower environments and thrive.

But the most astounding aspect of their morphology was undoubtedly the right/left asymmetry displayed by various species. As far as I know, no one has yet been able to explain why such a morphology could have established itself (if it is necessary to explain it!).

Stylophores, due to their bizarre and unusual anatomy, have been the subject of various taxonomic diatribes: those who considered these organisms as echinoderms, those who made them completely different animals… They were also considered to be close to chordates, the group that includes vertebrates, because of their appendix, which served as (a kind of) notochord.[22] In addition, the openings visible on the case of some forms have been considered as homologous with those of the oldest chordates. Finally, the discovery of the aquifer system has allowed them to be considered as real echinoderms!

Despite their ungainly appearance, these creatures have produced a dozen different species over more than 70 million years, from the Middle Cambrian to, at least, the Lower Devonian.

Having reviewed a bit of the morphology of the Palaeozoic echinoderms, what about their diversity during the entire era? If the number of species known during the Cambrian rises to 188, from 65 different localities, at the course of the Ordovician this value increases; no less than 1938 forms have been identified from more than 300 sites around the world.[23]

Of course, these values seem derisory to us compared to the total of modern forms; about 7000 species described (but, probably, many others are missing…). However, it should not be forgotten that the fossil data are only partial! For any given period of time, only a fraction of the species present has been able to fossilize and reach us. Some finds are simply incomplete and prevent us from properly assessing the biodiversity of the time. In addition, not all environments are able to fossilize the remains. Let us not forget that echinoderms are not all equally likely to be preserved; some, due to the particular composition of their tissues[24] or the environments frequented, are rarer than others, in the fossil state. Finally, most of the geological layers are no

[21] Depository-feeding creatures are those that feed on organic matter (organized in small particles) accumulated on the bottom of the seas and oceans.

[22] The notochord is a cartilaginous structure located in the dorsal part of the embryo of certain animals. All those who possess it belong to the group of chordates. In one subgroup, the vertebrates, it transforms into the spinal column.

[23] My information on the echinoderm biodiversity during the Cambrian is derived from the article by Zamora et al. (2013). Those relating to the Ordovician are from Lefebvre et al. (2013).

[24] For example, sea cucumbers have transformed calcareous plates into simple, more discreet spikes. They have lost in protection but have gained in mobility. We know about 900 fossil species, from the appearance of the group, up to the Pleistocene. The modern forms, on the other hand, are no

longer accessible to us, because due to plate tectonics, they would have been destroyed, dragged into the bowels of the Earth.

The number of Ordovician echinoderms is impressive. It is in fact during this period that all the current groups appear, sea urchins, crinoids, and also sea cucumbers, starfish and ophiuroids. But the five modern classes were not alone; they shared the rooms with a whole host of exclusively fossil groups, some of which were older than themselves. In some localities of the Upper Ordovician up to 14 different classes of echinoderms could live next to each other! I remember that nowadays, we can only find five at most, all present in the same location.

Echinoderm communities did not remain static during the Primary. Their structure has evolved over time. Some groups, exclusively Palaeozoic, began to lose importance in favour of faunae of different composition. Others, on the contrary, have multiplied. For example, during the Upper Ordovician crinoids become, in various localities, the most common echinoderms in terms of the number of species and/or individuals. Imagine an array of organisms, standing out like flowers, more or less compact, starting from the seabed, all intent on filtering suspended food particles from the water. It is nothing more than an image of a "prairie" composed of Palaeozoic crinoids. In addition to sea lilies, the environment is also frequented by other pedunculated echinoderms (for example, rhombiferans).

Crinoids continue to make up a substantial part of echinoderm faunae even during later periods. At Hunsrück Slate (Early Devonian, ~ 390 My), in the Rhine Massif (Germany), it is possible to identify about 60 different species of sea lilies,[25] number comparable, or rather, higher, than that of most modern localities. The fossil forms could be differentiated from the morphology and number of the arms and from the particular characteristics of the calyx and peduncle. They shared the environment with numerous starfish, sea cucumbers and other echinoderms known only in the Primary Age.

But the Palaeozoic echinoderms are not only of interest to us as elements of the biodiversity of the time. Their fossils provide us with irrefutable evidence of interactions between different species more than those provided by predation. Leaving aside the case of organisms that used the stems of sessile echinoderms as a support to encrust themselves and rise above the seafloor (or

less than 1430. One of the reasons why sea cucumbers are relatively rare and their fossils are usually isolated spikes, and that special collection methods are often needed to find them.

[25] An excellent book on fossil, Palaeozoic and Mesozoic crinoids and their diversity is by Hess and his colleagues (2003). For those who can read French, I also recommend Lebrun (2011) and Mirantsev (2012). Although their work is limited to the Carboniferous, both articles offer good examples of the diversity of Palaeozoic crinoids.

simply to find a rigid substrate to adhere to), it is possible to observe much more convincing evidence. For example, on the peduncles of some crinoids found in Ordovician sediments near Cincinnati (Ohio, United States) it is possible to notice bulges similar to "galls".[26] Similar structures are also found on the stems of modern pedunculated crinoids. Normally, they are produced by myzostomid worms, sea creatures that are often considered neighbours annelids.[27] These animals enter the tissues of echinoderms to find shelter and nutrition. Despite some differences found between modern and fossil structures, there is no shortage of specialists who attribute the Ordovician "galls" to the work of Palaeozoic myzostomids.

These are not the only close interactions between sea lilies and other creatures. In fact, we know of various fossil crinoids with a small gastropod installed on the echinoderm calyx. The mollusc is installed right above the anal pyramid of the sea lily (Fig. 4.4). We do not know if the gastropod was a parasite (like molluscs belonging to the genera *Thyca* and *Ophyotrix*, which mainly parasitize echinoderms) or if it is a simple relationship of commensalism.[28]

As mentioned earlier, it is clear that echinoderms, from their first appearance, began to establish a series of interactions with contemporary organisms similar to those still observed in modern forms. Palaeozoic echinoderms were true protagonists of the biodiversity of their time. But they were not alone…

4.3 "Continental" Stories and Plant-Insect Interactions

Having explored the aquatic environment, let us now turn to the continents to examine their biodiversity. Along the course of our study, we have already encountered organisms that left the marine environment. Some lived in freshwater[29] (thus continental) habitats, while others may have been able

[26] The galls, in botany, are growths produced on the stems, leaves or fruits of certain plants, due to bites of parasitic animals (especially insects). By analogy, the term is used to refer to a whole series of bulges found in the animal world. Regarding the fossil evidence of such "accidents" on the Paleozoic crinoids, see, for example, Meyer et al. (2009, Chap. 12).

[27] Annelids are worm-like animals whose bodies consist of a more or less high number of similar segments. Most annelids live in the sea; however, the group also includes terrestrial forms such as earthworms.

[28] Commensalism is a biological interaction between two different species. One species, the host, provides food (or protection) to the other, the commensal, without receiving any obvious compensation in return. It is a non-parasitic form of "exploitation" in which one species benefits from the other without mutual gain.

[29] See paragraph 3.3.

Fig. 4.4 Reconstruction of the Ordovician crinoid *Glyptocrinus decadactylus* with a gastropod of the genus *Cyclonema* attached to the calyx

to survive almost entirely without water.[30] Nevertheless, various lines of evidence indicate that during the Palaeozoic, atmospheric oxygen levels were high enough to allow organisms to venture onto land.

Let us begin with the plant world. The oldest traces of land plants, or at least of organisms assumed to be land plants, are spores found in Ordovician sediments, over 470 million years old. In contrast, macrofossils (as remnants of stems, roots, or leaves) appear in the following period, the Silurian, which began around 440 million years ago.[31] From this period onward, plants began to differentiate in terrestrial environments; by the end of the Devonian (~ 375 million years ago), lush forests lined the edges of swamps.[32]

During the Palaeozoic, plants achieved a major evolutionary breakthrough to establish permanently on land: the seed. This new structure facilitated dispersal and allowed plants to germinate far from their parent organisms. Compared to spores, seeds better protected the plant embryo and, having

---

[30] It is thought that some vendobionts, those organisms typical of the Eudiacarian fauna, may have lived in terrestrial environments like lichens. However, not all specialists disagree on this point. Let us go to the end of paragraph 3.5 and footnote 47.

[31] For the oldest known traces of terrestrial plants, see Taylor et al. (2009, Chaps. 6 and 8).

[32] Taylor et al. (2009, Chap. 12) and Steyer (2009). These works concern the fossil forms. All those who are passionate about trees and plants can read with pleasure the book by Wohlleben (2016).

stored reserves before germination, enabled development even under initially suboptimal conditions.[33] All major groups of terrestrial plants appeared during the Palaeozoic, except for flowering plants.

Regarding animals, when did they leave aquatic environments? The oldest evidence comes from a series of footprints preserved on terrestrial surfaces dating to around 485 My, at the Cambrian–Ordovician boundary (MacNaughton et al. 2002), in present-day Canada.

Which animals have left these footprints? These traces were apparently made by various arthropods, all walking on exposed land surfaces. The cuticle (the external skeleton covering their bodies) must have offered sufficient protection against harmful radiation reaching the soil.

However, it remains unclear whether these animals were primarily aquatic, occasionally venturing onto land, or truly terrestrial, capable of permanent life on land. Complete fossil remains would be needed to resolve this. Nevertheless, it is certain that at least by 480 My, some arthropods could leave water for short periods, a date preceding all known plant evidence. Unlike animals, plants cannot temporarily abandon their environment[34]!

After these first steps, the achievements of the new environment become more and more numerous. By the Lower Devonian, around 410 My, one of the oldest known fossil ecosystems developed in Scotland, near the village of Rhynie. The site, exploited for its shales, allowed the discovery of a community made up of plants, animals, fungi and, of course, microbes. There is no shortage of examples of interaction between plants and fungi, with associations interpreted as beneficial for the two partners and others that, on the other hand, testify to real cases of parasitism towards the plant.[35] In short, the Rhynie ecosystem presents us with fossil examples of those same interactions that are currently contracted between these two types of living beings. Not only that, they also show us that the interdependence between plants

[33] Of course, sooner or later, the conditions must become compatible with the development of the embryo before all the resources available to the semen have been exhausted, otherwise growth will stop and the organism will decay. The seed is a protective cover for the plant, including a survival kit which can be used in difficult conditions. It is not an asset that can magically solve all critical situations. On the other hand, if the plant bets on the stability of the environment and its. ability to allow germination in all situations, the seed is not indispensable, as evidenced by the existence, nowadays, of plants that still entrust their reproduction to spores (for example, ferns). However, that the seed was one of the keys to promote the colonization of the land is proved by the vast majority of terrestrial plants that use it. This is the case for all conifers and flowering plants.

[34] Of course, it is not always possible to differentiate the animal traces left by amphibious creatures, which move on land only occasionally, from those of organisms that, instead, reside there permanently. With plants, fixed in place, doubt arises more rarely. Having said this, the progress of research (new discoveries, new means of investigation to interpret already known data) may reserve us surprises in the future regarding the exact timing of the "conquest" of the land by the different organisms.

[35] For the fossil fungi and lichens of the Rynie schists, see Taylor et al. (1995, 1999). Regarding interactions between plants and fungi, see Taylor et al. (2009, Chap. 3), and Selosse (2017, Chap. 3).

and fungi seems to have been essential to facilitate the establishment of the former in terrestrial environments.

Why did organisms leave water for a new and hostile environment? What motivations may have driven them? The question seems rhetorical: it is from their appearance that living beings settle in every environment as soon as it becomes "habitable"! To escape predators, to find new resources, to avoid competition with other creatures. The reasons are varied and diverse.

The so-called "pioneer" species established conditions, in the new environments, that allowed subsequent species to settle.[36] Soon, therefore, a whole group of animals (arthropods and vertebrates[37]) settles on the continents.[38]

In the rest of the paragraph, I want to focus on the group of insects, simply to give you some information about them and to describe the interactions with the plants of the time.

Insects today represent the largest group of terrestrial organisms, inhabiting nearly every terrestrial and freshwater environment. Marine insects are rare, with crustaceans fulfilling many ecological roles. The earliest insect fossils date to the Devonian, with major diversification during the Carboniferous and Permian. By the late Palaeozoic, the four principal insect groups had already appeared.[39]

---

[36] It is classical, for example, the colonization of a new island, which arose from the sea due to volcanism. The first to settle are plants whose seeds have arrived on site, along with the seabirds that nest and feed in the waters around. Later, arthropods may arrive, carried by the wind or using other means of transport. Finally, other organisms arrive. In short, some species arrive earlier because they are more adaptable. These modify the environment in order to make it more "comfortable" for newcomers.

[37] When talking about those animals that have invaded the land, people invariably think of arthropods (insects, spiders, scorpions, millipedes) and vertebrates. However, many other groups have been able to leave the liquid environment. Gastropods, for example, which, among snails, snails and the like, count something like 25,000–30,000 different modern species; almost as much as those of bony fish (approximately 32,000 modern species)! However, the history of their colonization of the land is not as well-known as that of arthropods or vertebrates (or, it is much less publicised). Some information can be found in Stworzewicz et al. (2009) and Mordan and Wade (2008). Instead, I confess that I have not been able to find out anything about the evolutionary history of earthworms and all those parasites that infest terrestrial organisms.

[38] A good popularization text on the "conquests" of terrestrial environments by animals is that published under the direction of Gould et al. (1993). The history of adaptations that allowed vertebrates to settle on the land and the evolution of the tetrapod limbs (this is the name given to the group to which terrestrial, current and fossil vertebrates belong) is masterfully told by Steyer (2009).

[39] The four groups are: the apterians (without wings), the palaeopterans and the neopterans (respectively, the insects unable, at rest, to fold the wings on their abdomen and those who can do so), the holometabola (the insects that pass through a complete metamorphosis between the larval stage and the adult—and in which the larva is very different, morphologically, from the adult individual). For the oldest known insect fossils and the evolutionary history of the group, see the article by Garrouste et al. (2012) and the book by Grimaldi and Engel (2005).

The most common Carboniferous–Permian insects were palaeodictyopterans. With piercing mouthparts, they were primary herbivores. Some had lateral expansions on the first thoracic segment resembling small wings (Fig. 4.5), earning them the nickname "three-paired-wing" insects. The best-known insect of the Palaeozoic is undoubtedly the giant dragonfly *Meganeura* (Fig. 4.6). With a wingspan of almost 70 cm, it was the size of a hawk.[40] But *Meganeura* wasn't the only giant arthropod at the time. *Arthropleura*, for example, was a kind of prehistoric centipede almost as long as a cow (Grimaldi and Engel 2005; Schneider and Werneburg 1998)! A number of palaeodictyoptera, without being colossal, were respectably large, being in the upper half of those of ordinary insects. Was the Carboniferous-Permian therefore an era of giant arthropods? And why could insects (for example) grow like this? The most commonly used explanation tells us that, at the time, the oxygen content in the atmosphere was higher than it is today.

Insects breathe through the diffusion of air through their body. The gas penetrates through openings in the chest and abdomen; it then spreads into the animal's body thanks to a system of tracheae. The latter are made up of simple empty tubes that allow air (and, therefore, oxygen) to pour into

Fig. 4.5 Reconstructions of Palaeozoic insects: on the left, the paleodiptyoptera *Stenodictya lobata*, found in the upper Carboniferous strata of Commentry (Allier, France); on the right, one of the beetles' ancestors, *Permocupoides sojanensis*, which lived during the Permian. The two insects are not represented at the same scale

40 *Meganeura monyi*, from the Carboniferous of Commentry (Allier, France), was considered, for some time, as the great insect known. However, the record is currently held by the American species *Meganeuropsis permiana*, whose wingspan measures about 710 mm! *Meganeuropsis* belonged to the same group as *Meganeura* (Grimaldi and Engel 2005).

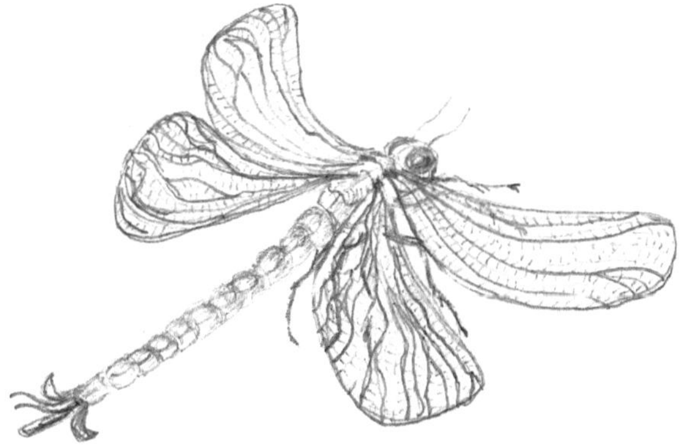

Fig. 4.6 *Meganeura monyi*, the Carboniferous giant dragonfly, with a wingspan of almost 70 cm (Commentry, Allier, France)

every organ and every part of the body. Oxygen movements are carried out mainly by simple diffusion, although, in some cases, there may be pockets with the function of "pumping air". This system is significantly less efficient than that available to vertebrates. Diffusion is effective only in small bodies, which explains why insects are generally not gigantic.[41] A fly the size of a cow would exceed the capacity of this system to supply sufficient oxygen.

A higher oxygen content, compared to the modern value, therefore seems to be able to explain the phenomenon. But do we have any additional evidence, apart from giant insects, to assess the amount of O_2 present in the atmosphere in this time? What do Carboniferous-Permian plants tell us about this? In fact, the amount of oxygen present in the air leaves traces on the functioning of plant processes, especially on the fractionation of stable isotopes of carbon.[42] Thus, chemical analysis of the fossil organic remains of Carboniferous-Permian plants can help us assess the oxygen content of the time.[43] The results obtained would be in agreement with a quantity of

[41] One must be careful with generalizations; even today there are large insects. The heaviest insect is an African beetle, the Goliath beetle (genus *Goliathus*). It weighs about 100 g. Another giant is the Malaysian phasmid (*Heteropteryx dilatata*). The female (the females of these insects are larger than the males) can be over 15 cm long and weigh up to 70 g. However, these animals are likely to be much less active than the Carboniferous-Permian giant dragonflies and the anatomical devices described in the text are fully sufficient to enable them to carry out their daily activities.

[42] Do you remember carbon isotopes? We have already discussed this in Sect. 2.1.

[43] In fact, a key enzyme involved in the photosynthetic phases of CO_2 fixation and glucose production (remember that photosynthesis allows plants to produce food—glucose sugar—on their own?) is sensitive to the oxygen content in the air. This molecule is called RubisCO and in plants it consists of four monomers (sub-units), like the haemoglobin in our blood. Just like haemoglobin, the RubisCO enzyme can cope with CO_2 as well as oxygen. The content of this gas in the air can alter the function of the enzyme as well as the isotopic fractionation of carbon fixed during photosynthesis. That is why chemical analyses of organic carbon in plants can tell us about the amount of oxygen in the

oxygen in the air, towards the end of the Primary, which could reach 35% (against the current 21%).

The idea is serious, but things may have been even more complex! In fact, our data on Carboniferous-Permian insects do not appear to be exhaustive. The fossil record of the latter part of the Palaeozoic offers not only giant species, but also small forms. The insects were not all large. The questions to ask ourselves are therefore other. During the Carboniferous and Permian, was the ratio of "large" to "small" insects much higher than it is today, or not? Is the result of the Palaeozoic value the result of an artifact of fossilization (larger insects may be better preserved in sediments than smaller ones, or their fossils may simply be easier to find) or is it a real phenomenon? Despite the data accumulated to date, we are still unable to definitively answer the questions. For example, recently, an international team of palaeontologists discovered a French fauna of small Palaeozoic insects, much smaller than those of the animals previously evoked. The new species described do not belong to exclusively fossil groups, such as giant "dragonflies" and paleodiptyoptera. These are some of the oldest representatives of the groups that include, among others, ladybugs, flies, bees and butterflies (Nel et al. 2013).

Do these new species oppose a higher oxygen content during the Carboniferous and Permian than at present? Absolutely not, because if a high oxygen content is important to ensure the existence of colossal insects, it does not prevent that of small arthropods at all. However, their presence makes us reconsider the causes of insect gigantism, because this now seems limited to particular groups. The causes of the phenomenon could therefore be more complex than those invoked to date. It could be a series of palaeoecological, physiological and environmental factors, in which the amount of oxygen present in the atmosphere is only part of the answer (perhaps not even the most important!). The suspense continues…

Despite all the talk about the size of Palaeozoic insects, they profoundly influenced plants, and vice versa. For example, *Blazyopteris praedentata* a seed-producing fern[44] growing, similar to a vine, in the European forests of the Late Carboniferous, had peculiar features. The lower surface of its fossil fronds has growths (called, in technical terms, "trichomes") of the plant epidermis with various configurations. These can be straight or curved, in the shape of a hook. The trichomes, in modern plants, act as deterrent against vegetarian arthropods. Other trichomes, of different shapes, are interpreted as

atmosphere while they were active. For the full history of oxygen content during the Carboniferous and Permian periods, see Beerling (2008, Chap. 3) and McGhee (2018).

[44] The "seed ferns" are a group of plants similar, in morphology, to ferns but that were reproduced thanks to seeds. Todays, there are no longer representatives of this group.

glands capable of providing more effective defences against herbivores (odours or other repulsive chemicals), always in line with the defences developed by modern plants (Krings et al. 2003). Our seed fern was not a passive victim to the attacks of the herbivores of the time!

Insects have also been inspired by plants for their evolution. We are well aware of the cases of camouflage between the wings of certain insects (cockroaches) and the Palaeozoic leaves. These fossils have been found in Carboniferous and Permian sediments. Specialists think that insects residing on specific fronds had become able to imitate, thanks to the design of the veins on their wings, the decorations of the leaves (Taylor et al. 2009, Chap. 23). Probably, so as not to be found by predators (for example, by giant dragonflies).

However, the most famous interactions between insects and plants are, of course, those related to pollination. But flowering plants did not yet exist! According to fossil evidence, these plants will only appear in the Mesozoic (we will deal with them in the next chapter). But flowering plants are not the only ones to be fertilized with the help of insects.

Cycads[45] form a set of modern (Fig. 4.7) and fossil plants, the oldest known representatives of which date back to the Late Carboniferous, starting from about 320 My. The pollination of many modern species (probably all of them) is ensured by insects (Taylor et al. 2009, Chap. 17), especially beetles and even some thrips (Thysanoptera). The latter[46] are well known for interacting with cycads. Among other things, they feed on their pollen. However, in Australia, four species of cycads, belonging to the genus *Macrozamia*, depend on thrips of the genus *Cycadothrips* (Fig. 4.8) for reproduction (Terry 2002).

Is it conceivable that fossil cycads were pollinated by insects? We have indirect evidence supporting this possibility. In fact, on a pollen cone of a *Delemaya spinulosa* specimen found in the Antarctic, in Middle Triassic sediments (between 247 and 237 My, thus post-Palaeozoic), coprolites (fossil excrement) containing exclusively pollen grains of the plant species in question have been identified (Klavins et al. 2005). This discovery, if interpreted correctly, would constitute the oldest evidence of plant pollination by an

[45] Cycads are palm-shaped gymnosperms (plants in which the seed is bare—like pine, fir and cypress—opposite to angiosperms, flowering plants). Their morphology would have changed very little in the course of the evolution of the group. The breeding structures are made up of cones located on top of the plant. Cycads have different sexes: there are males, which produce pollen, and females, which produce eggs.

[46] The thrips (Thysanoptera) are very small insects (the gold size is often less than 2 mm) characterized by the morphology of the wings. Their veining system is reduced and, above all, they have long silk fringes on the edge of the wings. These "filaments" help to increase the supporting surface of the wing. Fossils of their group were found in the Upper Carboniferous of France (Nel et al. 2012).

Fig. 4.7 Japanese cycads (*Cycas revoluta*), a plant native to the Ryūkyū and Nausei islands, located between Japan and Taiwan

Fig. 4.8 Thrips of the genus *Cycadothrips* pollinating the Australian cycads

insect (assumed to be the animal that excreted the pollen), predating by over 100 My the oldest known flowering plant fossil!

Could cycads have been pollinated as early as the Primary Era? This is entirely plausible, as these plants had already appeared by that time. Regarding insects, the group including thrips (Thysanoptera) is known from the Upper Carboniferous, and the first beetles date from the end of the Palaeozoic[47] (Fig. 4.5). What is certain is that plant-insect interactions were

[47] For the oldest known fossil thysanoptera see Nel et al. (2012). For beetles, I recommend to consult the book by Grimaldi and Engel (2005) and Nel et al. (2013).

already well established by the late Primary Era, during the Carboniferous and Permian periods, and subsequently became increasingly complex.

4.4 The End of a World

The Primary Era ends around 252 My. If its beginning is characterized by an explosion of Life,[48] a reflection of the exceptional number of fossils found in Palaeozoic sediments compared to the Precambrian, its end is marked by the most significant mass extinction on record.

What is a mass extinction? This concept, introduced in paragraph 3.4 regarding a possible crisis triggered by oxygen diffusion from Precambrian photosynthetic microorganisms, can be defined more precisely as a drastic reduction in biodiversity (a significant loss of species) over a relatively short geological period.[49] This decline must affect numerous organisms differing in size, morphology, taxonomy[50] and even habitat, over a vast geographic area including multiple continents and marine zones: a global catastrophe! The greater the biodiversity loss, the more severe the mass extinction.

While the Precambrian crisis mentioned earlier remains speculative due to insufficient evidence, later extinctions are well documented. Throughout Earth's history, five major crises have been recognized, notable for the sheer number of species lost (Raup 1991). These are not only the extinction episodes that take place on Earth, but those that have claimed the highest number of victims, among all those known. Here is the list, in chronological order:

- the first is located towards the end of the Ordovician (~ 444 My);
- the second, in the terminal phase of the Devonian (primarily between the Frasnian and Famennian ages, ~ 372 My, although it was probably a complex phenomenon, which lasted until the end of the era);
- the third, at the end of the Permian (~ 252 My);
- the fourth, at the time of the Triassic/Jurassic transition (~ 201 My);

---

48 In reality, the beginning of the Palaeozoic may have coincided with a true biological crisis. In fact, if it is true that several previously unknown life forms appear, we also witness the extinction of all (or almost) the Ediacarian creatures. In this respect, see Gould (2000).

49 When we talk about biological crises, the time interval of the crisis is never (or rarely) defined. In fact, this time is difficult to estimate. It is a negligible amount of time on the part of the time scale that we are interested In. For example, for the Palaeozoic crises, it can last up to a few million years (or even more). The most recent periods, such as the set of extinctions at the end of the Pleistocene, are clearly shorter: a few thousand years at the most.

50 Taxonomy is the science that identifies and describes living beings to classify them into systematic groups according to their relationships.

- the fifth, Cretaceous/Tertiary (~ 66 My), is placed at the transition from the Mesozoic to the following era, the Cenozoic.

The last episode is the most famous for the extinction of the dinosaurs, while the third (the Permian/Triassic) is the deadliest.

The first major extinction drastically reduced trilobites (losing two-thirds of their families) and conodonts[51] (losing 80% of species). Other animal communities, such as molluscs, are more resilient to the crisis.

At the Frasnian/Famennian boundary, all jawless armoured vertebrates disappeared,[52] along with many trilobites (which had partly recovered from the previous crisis) and other marine organisms, especially those that lived in shallow waters.

At the end of the Permian crisis, about 90% of marine species disappeared; on land the situation went a little better, as losses were limited to 70%!

At the Triassic/Jurassic boundary, many terrestrial vertebrates and over 50% of marine species went extinct.

The K/T extinction The Cretaceous/Tertiary transition (often abbreviated as K/T), saw roughly 70% of species perish, including all non-avian dinosaurs,[53] flying reptiles, various mammals (notably marsupials), and numerous birds.

At the transition between the Permian and Triassic, 90% of the Marine Life mansion disappears forever. But what exactly does this percentage mean? What exactly does it represent? In order to calculate it, it is first necessary to list all the living beings that have lived just before the limit sanctioned by the crisis and just after, and then compare the two values. The only way to account for these organisms is to do so from their fossil traces. The phrase "90% of life forms have become extinct" must be translated into "90% of life forms, KNOWN FROM FOSSILS, have become extinct".

But the evidence left by the fossil traces is necessarily incomplete. All organisms with highly biomineralized tissues or living in environments that favour fossilization (environments where organic remains are rapidly buried and/or oxygen-free environments) will be more easily preserved. On the

[51] The term "conodont" refers to tiny fossils, consisting of phosphates, having the form of small teeth. The name, by extension, has now also designated the animals that owned these facilities. These were located in the front of the body, which had a vermiform shape. About these bodies see Blieck (2012).

[52] Vertebrates without a jaw are called "agnates". To this group belong the lampreys and the present myxines, more different fossil groups, exclusively Palaeozoic. For further information on the agnates, see Janvier (1996) and Long (2011).

[53] As it is now established that birds are the legitimate descendants of large Mesozoic reptiles, the term "non-avian dinosaurs" is increasingly being used to differentiate the first from the real dinosaurs.

contrary, those who have a soft body, without rigid tissues, or who frequent environments with a low rate of fossilization, risk never being known by palaeontologists. Earthworms, for example, are practically unknown in the fossil record, while they should have been present from ancient times…

Despite these limitations, the accumulated data allow reliable conclusions. The percentages provided above may be modified following new data, but the general picture will remain unchanged. Biological crises will continue to mean real hecatombs, and the Permian/Triassic crisis is unlikely to lose its deadliest extinction status on record.

Having said that, let us return to the crisis that interests us to examine, in more detail, the victims, the survivors and the causes of the die-off. The most famous victims are undoubtedly trilobites. This group, whose diversity was, in fact, already weakened by two different extinction episodes, became completely extinct at the end of the Palaeozoic. But they are not the only ones to suffer this fate. In the seas, of the four groups of fish present (armoured fish, or placoderms; spiny "sharks", or acanthodes, cartilaginous fishes [= sharks and related species] and bony fishes) only two cross the Permian/Triassic limit,[54] and not without having suffered a certain number of losses. From the survivors, through an evolution that will last throughout the Mesozoic, modern forms will originate. All the coral groups typical of the Primary also vanish. Among the echinoderms, only five classes see the dawn of the new age dawn; five classes whose ranks have been decimated. And the same is true of most other marine animal groups. But the crisis does not only affect animals; even at the level of microorganisms there are significant losses. Take, for example, a group of foraminifera,[55] the fusulinidae. There will be no trace of these creatures since the Triassic!

On land, insects lost around eight groups (about two-thirds of the forms known at the time), including giant dragonflies and palaeodictyoptera, evoked in paragraph 4.3. And if plants were able to overcome the crisis with limited losses, the same cannot be said for terrestrial vertebrates; they lost about two-thirds of their populations were permanently eradicated from the biodiversity of the time. Among these, the reptiles called "mammalians"

[54] It should be noted, however, that the armoured fish had already become extinct at the end of the Devonian, more or less 359 My ago. These vertebrates have survived the beginning of the Frasnian/Famennian crisis, which occurred a few million years earlier, but none of them seems to have reached the Carboniferous. So, the only group of fish to really become extinct at the end of the Permian is that of the acanthodes.

[55] Foraminifera is a group of marine micro-organisms capable of secreting a tiny shell around their body, the shape of which is typical for each species. The shell of foraminiferous (except for some species) is made of limestone; the organism lives inside but can make out the expansions of the body to capture the food it eats (Decrouez et al. 1996). Foraminifera are an old group, which appeared in the early part of the Palaeozoic.

(because it is within this group that the ancestors of mammals must be sought), were in full expansion; but they were hit hard. However, two different evolutionary lines were able to survive and continued their evolution during the Triassic.

The result that emerges from the picture described is that the victims are all different in terms of taxonomic belonging, in the environment frequented and in geographical origin. And even if we consider that the crisis, at least in the seas, could have occurred in two successive waves of extinctions, separated by at least 200,000 years,[56] this does not detract from the seriousness of the event, nor from its rapidity.

What cause, or more precisely, what causes, caused such a catastrophe? The debate is still open, but various avenues have been identified.[57] First of all, it must be pointed out that, at the time, the continental masses were united in a single super-continent: the famous Pangaea. This configuration had the effect of considerably reducing the marine environments of the continental shelf (the coastal environments of shallow depth) of the time. These are, in fact, all those different environments that are found in front of the coasts, just before the rapid underwater slopes that overlook the abyssal bottoms. If the length of the coasts decreases, these environments shrink in turn!

Of course, the meeting of continents did not take place in a day. The organisms typical of such environments have had, in this case, time to adapt to the new situation. But if we add to this stress other perturbations (for example, changes in sea level), the inhabitants of shallow basins find themselves subjected to constraints that endanger their existence (is it worth remembering that the extinctions at the Permian/Triassic limit mainly involved sea creatures?).

But there is more; the reunion of the continents in a single large mass also has profound effects on the climate. In fact, it favours the formation of vast desert expanses, in the central part of the continental block, with an increase in aridity and, consequently, a decrease in vegetation cover. Vegetarian organisms must then concentrate on the coasts where humidity allows plants to grow. However, the rarefaction of these environments exacerbates competition and can cause extinctions. All this punctually occurred during the Upper Permian!

---

[56] An international team, thanks to its work, would have highlighted that, at least in the Chinese region where fossil collections were concentrated, the Permian-Triassic extinction was not a specific event. The researchers have identified at least two different peaks of disappearance: the first one ~ 100,000 years before the end of the Permian, the second one ~ 100,000 years after the beginning of the Triassic (Song et al. 2013).

[57] For more details on the Permian-Triassic mass extinction and its possible causes, see Steyer (2009, Chap. 4, pp. 159–162), and Crasquin et al. (1989).

However, among the possible causes of this crisis, one more than the others have considered the attention of specialists, at least in recent times: that of increased volcanic activity. According to this hypothesis, volcanoes would have emitted a huge amount of harmful gases in the form of clouds at the end of the Permian. One of the effects would have been to cause a decrease in sunlight on Earth, with harmful consequences for plants, but also causing considerable losses at all levels of animal life. To corroborate these hypotheses, the scientists cite the Siberian traps, lava flows that extend over immense surfaces (about 200,000 km^2) and correspond to massive volcanic episodes concentrated at the end of the Permian.

And what about the fall of a meteorite? If a celestial body that struck our planet was implicated in the K/T crisis, why couldn't a similar event have marked the end of the Palaeozoic? This eventuality could also explain the (presumed?) regularity of the extinction phenomena reported by some scientists (Raup 1991).

The meteorite hypothesis is therefore seductive, but what is the evidence? For now, they are not very numerous. No one has ever found the impact crater yet. However, the research is only beginning and it is not impossible that an area with the required characteristics may not one day be highlighted. Furthermore, the fall of meteorites or asteroids must not have been uncommon in the past, it is judged, for example, at the lunar surface.

So, who is to blame for the Permian/Triassic crisis? The fall of a meteorite? Volcanic activity? The reunion of the continents in a single mass? Given the complexity of the event (a massacre of disparate living beings, on land and at sea), it is very likely that various causes are involved. It is possible that marine eustatic movements,[58] in a context of reduction of shallow coastal environments, have made themselves felt on the typical organisms of such environments. If we add to this the pollution due to significant emissions of CO_2, a gas that is preferentially absorbed by the seas and oceans, it is clear that marine communities have been tested hard.

On land, things did not go much better, given the climatic changes caused by volcanic phenomena and those related to the position of the continents, not to mention the harmful gases spread in the atmosphere. And if we also add the fall of one or more meteorites (which may have ended up indifferently on land or in the seas) here the picture becomes truly apocalyptic. The disappearance of a large number of animals should no longer make any mystery!

[58] An eustatic movement is a generalized rise or fall (recorded across the globe) of sea level relative to the coastline of continents.

However, this picture is only a guess. Other data are needed, especially those relating to one or more meteorite impact craters, to produce really satisfactory hypotheses on the causes of the Permian-Triassic crisis. Only the progress of scientific research will one day be able to give us the answer.

I want to end the paragraph by pointing out that, despite all that has just been said, biological crises should not be considered only as catastrophic events. The negative aspect is only one side of the coin; the other is more constructive for evolution. While they eliminate dominant groups, they create ecological opportunities for other organisms. After the mass extinctions have made a clean sweep, they have the ability to differentiate themselves, to link associations with other organisms, to show themselves no less efficient than all those beings that have just disappeared.

Let us take an example. Lawyer X is very famous, a true prince of the bar. His resolute and aggressive character won him numerous cases in the courts and made him in great demand. He is considered the best lawyer on the market! However, one night, following a party in which he drank a little too much, he has a car accident that is fatal to him.

His assistant, the lawyer Y, much shyer and more introverted than her boss, is forced to resume all the cases left unresolved. Her competence and tenacity, however, enable her to perform all tasks positively. No longer in the shadow of lawyer X, it is her method that imposes itself in the courts. Our lawyer Y becomes the new "queen" of the bar, even more requested and effective than her predecessor.

This is how a perfectly fortuitous accident (who would have thought that the lawyer would die that night?) brought out a high personality, no less valid than the one that preceded her, capable of "taking the baton" (as in a relay race) and taking it further!

Biological crises can therefore favour variety; after the domination of one or more groups, which monopolize biodiversity, comes a mass extinction that puts the cursor back to zero and offers a *chance* to other organisms, often very different from the first, which otherwise would not have had the opportunity to differentiate further. Following the Permian extinction, Triassic organisms (animals, plants and even microorganisms) gradually filled these voids, leading to increased ecological and evolutionary variety.

4.5 Summary of This Chapter

The early Palaeozoic period is characterized by the phenomenon known as the "Cambrian explosion of Life." This term refers to the sudden appearance of a large number of organisms, particularly animals with bilateral symmetry (all grouped together as "bilaterians"). The event is evidenced by the abundant fossils found in Palaeozoic sediments, clearly outnumbering the remains exhumed from Precambrian layers. However, this abundance may reflect a different phenomenon. Since most of these remains consist of biomineralized structures (which fossilize more easily than soft tissues) the large number of Palaeozoic fossils could simply result from organisms acquiring the ability to produce biominerals, rather than from a true diversification of life forms.

The causes of this phenomenon were likely multiple. On one hand, changes in ocean chemistry, especially an increase in oxygen content, counteracted water acidity caused by CO_2, favouring the precipitation of calcium carbonate and thus the formation of mineralized tissues. On the other hand, the evolution of more efficient organs (such as eyes) and/or gene complexes influencing embryonic development may also have played a role.

In any case, the difference from earlier faunas is evident. Palaeozoic seas and oceans sheltered an abundance of previously unknown animal forms (echinoderms, for example, diversified into 21 classes, 14 of which could coexist in the same location, compared to the five present today) including the first large predators.

On the continents, organisms capable of freeing themselves from aquatic constraints appeared, including arthropods, gastropods, vertebrates, and plants. Plants achieved most of their major evolutionary adaptations for terrestrial life, except for the production of flowers. In Carboniferous-Permian forests, the first interactions between insects, in full expansion, and plants were established.

The end of the Palaeozoic is marked by the most severe documented biological crisis in Earth's history; roughly 90% of marine organisms, across multiple taxonomic groups and distributed globally, vanished forever. On land, the crisis was slightly less severe, yet about 70% of terrestrial life forms became extinct. The causes proposed for this catastrophic event are varied: the assembly of all land masses into the supercontinent Pangaea (reducing shallow marine environments and creating vast deserts), increased volcanic activity, and the possible impact of a meteorite (though no impact crater has yet been identified). At present, no single cause is considered definitive; it is likely that multiple factors combined to trigger the crisis marking the transition to the next era, the Mesozoic.

Appendix

Variation of Biomineralization in the Marine Environment

In marine biology, as well as in geology and geochemistry, the lysocline is defined as the upper level below which the solubility of limestone increases significantly. This means that when organisms die, all calcium carbonate coatings, such as shells, have a tendency to dissolve, until they are completely consumed. Instead, above the lysocline, they can fossilize and be preserved within the sediments (of course if they have not already been destroyed by other types of accidents, such as shocks due to currents or other forms of transport). The lysocline is often associated with the carbonate compensation depth, abbreviated with the acronym CCD (Carbonate Compensation Depth). This is a depth greater than that of the lysocline that marks the limit of instability of carbonates. Below its level, no carbonate can survive. The CCD is not constant, neither in time nor in space. It varies from one marine region to another, depending on the pressure, temperature and chemical composition of the waters. The acidity of the water is an important factor that influences the solubility of carbonates: the more acidic the waters, the more the carbonates will be dissolved. CO_2 increases acidity and promotes the dissolution of limestone. Oxygen, on the other hand, acts in the opposite way, counteracting the action of CO_2. The presence of oxygen and its increase favour the formation of calcium carbonate deposits and therefore the production of exoskeletons (external skeletons) by marine organisms. From this point of view, the oxygenation of deeper and deeper marine layers has been essential to promote the ability of aquatic animals to secrete rigid tissues.

References[59]

Beerling, D. 2008. *The emerald planet—How plants, changed earth's history*. Oxford University Press.

*Blieck, A. 2012. Des pseudo-dents: Pourquoi les conodontes ne sont pas des vertébrés. *Fossiles* 11: 52–57.

*Bonino, E., and C. Kier. 2010. *The back to the past museum guide to trilobite*. Bergamo: Casa Editrice Marna S.C.

[59] Works preceded by an asterisk can be read, more or less easily, by people with a non-professional skilling.

*Briggs, D.E.G., D.H. Ervin, and F.J. Collier. 1994. *The fossils of the Burgess Shale*. Washington: Smithsonian Institution Press.

Canfield, D.E., and A. Teske. 1996. Late Proterozoic rise in atmospheric oxygen concentration inferred from phylogenetic and sulphur-isotope studies. *Nature* 382: 127–132.

Canfield, D.E., S.W. Poulton, and G.M. Narbonne. 2007. Late-Neoproterozoic deep-ocean oxygenation and the rise of the animal life. *Science* 315: 92–95.

Chen, J.-Y., J.W. Schopf, D.J. Bottjer, C.-Y. Zhang, A.B. Kudryavtsev, A.B. Tripathi, X.-Q. Wang, Y.-H. Yang, X. Gao, and Y. Yang. 2007. Raman spectra of a Lower Cambrian ctenophore embryo from southwestern Shaanxi, China. *Proceedings of the National Academy of Sciences U.S.A.* 104 (15): 6289–6292.

*Crasquin, S., J.-S. Steyer, and J.-O. Baruch. 1989. La plus grande extinction du vivant. *La Recherche* 142: 48–55.

David, B., and R. Mooi. 1999. Comprendre les échinodermes: la contribution du modèle extraxial-axial. *Bulletin de la Société Géologique de France* 170 (1): 91–101.

Decrouez, D., J.-P. Debenay, and J. Pawlowski. 1996. *Les Foraminifères actuelles*. Paris: Masson.

Erwin, D.H., and J.W. Valentine. 2013. *The Cambrian explosion—The construction of animal biodiversity*. Greenwood Village (Colorado): Roberts and Company.

Fedonkin, M.A. 2003. The origin of the metazoa in the light of the Proterozoic fossil record. *Paleontological Research* 7 (1): 9–41.

Fortey, R. 2000. *Trilobite! Eyewitness to evolution*. London: Harper Collins.

Garrouste, R., G. Clément, P. Nel, M.S. Engel, P. Grandcolas, C. D'Haese, L. Lagebro, J. Denayer, P. Gueriau, P. Lafaite, S. Olive, C. Prestianni, and A. Nel. 2012. A complete insect from the Late Devonian period. *Nature* 488: 82–85.

*Gould, S.J. 2000. *Wonderful life—The Burgess Shale and the nature of history*. New York: Vintage.

*Gould, S.J., P. Andrews, M. Benton, C. Janis, J.J. Sepkoski, and C. Stringer. 1993: *The book of life*. London: Ebury Hutchinson.

Grimaldi, D., and M. Engel. 2005. *Evolution of the insects*. Cambridge: Cambridge University Press.

Hess, H., W.I. Ausiche, C.E. Brett, and M.J. Simms. 2003. *Fossil crinoids*. Cambridge: Cambridge University Press.

Hou, X.-G., R.J. Aldridge, J. Bergström, D.J. Siveter, D.J. Siveter, and X.-H. Feng. 2008. *The Cambrian fossils of Chengjiang, China—The flowering of early animal life*. Oxford: Blackwell Publishing.

Janvier, P. 1996. *Early vertebrates*, 33. Oxford: Clarendon Press.

Klavins, S.D., D.W. Kellogg, M. Krings, E.L. Taylor, and T.N. Taylor. 2005. Coprolites in Middle Triassic cycad pollen cone: Evidence for insect pollination in early cycads? *Evolutionary Ecology Research* 7: 479–488.

Krings, M., D.W. Kellogg, H. Kerp, and T.N. Taylor. 2003. Tricomes of the seed fern *Blanzyopteris praedentata*: Implications for plant-insect interactions in the Late Carboniferous. *Botanical Journal of the Linnean Society* 141: 133–149.

*Lebrun, P. 1995. *Trilobites*, hors-série 2. Minéraux and Fossiles.

*Lebrun, P. 2011. Les crinoïdes du Mississippien (Carbonifère) de Crawfordsville, Indiana, Etats-Unis. *Fossiles* 7: 21–41.

Lefebvre, B. 2003. Functional morphology of Stylophoran Echinoderms. *Palaeontology* 46 (3): 511–555.

Lefebvre, B. 2022. *La diversification des échinodermes au Paléozoïque inférieur: l'apport des gisements à préservation exceptionnelle*. Lyon: Dédale Éditions.

Lefebvre, B., C.D. Sumrall, R.A. Shroat-Lewis, M. Reich, G.D. Webster, A.W. Hunter, E. Nardin, S.V. Rozhnov, T.E. Guensburg, A. Touzeau, F. Noailles, and J. Sprinkle. 2013. Palaeobiogeography of Ordovician echinoderms. *Geological Society of London* 38: 173–198.

Levinton, J.S. 2008. The Cambrian explosion: How do we use the evidence? *BioScience* 58 (9): 855–864.

Levi-Setti, R. 1995. *Trilobites*, 2nd ed. The University of Chicago Press.

*Long, J.A. 2011. *The rise of fishes*, 2nd ed. Baltimore and London: Johns Hopkins University Press.

MacNaughton, R.B., J.M. Cole, R.W. Darlympe, S.J. Braddy, D.E.G. Briggs, and T.D. Lukie. 2002. First steps on land: Arthropod trackways in Cambrian-Ordovician eolian sandstone, southeastern Ontario, Canada. *Geology* 30 (5): 391–394.

Marshall, C.R. 2006. Explaining the Cambrian "explosion" of animals. *Annual Review of Earth and Planets Sciences* 34: 355–384.

*McGhee, G.R. Jr. 2018. *Carboniferous giants and mass extinction*. New York: Columbia University Press.

Meyer, D.L., R.A. Davis, and S.M. Holland. 2009. *A sea without fish—Life in the Ordovician sea of Cincinnati region*. Bloomington, Indianapolis: Indiana University Press.

Mirantsev, G.V. 2012. Les crinoïdes du Carbonifères supérieur des carrières de Myachkovo (Moscou, Russie). *Fossiles* 10: 44–47.

Mordan, P., and C. Wade. 2008. Heterobranchia II—The Pulmonata. In *Phylogeny and evolution of the Mollusca*, ed. W.F. Ponder and D.R. Lindebrg, 409–426. Berkeley: University of California Press.

Nel, P., D. Azar, J. Prokop, P. Roques, G. Hodebert, and A. Nel. 2012. From Carboniferous to recent: Wing venation enlightens evolution of thysanopteran lineage. *Journal of Systematic Palaeontology* 10 (2): 385–399.

Nel, A., P. Roques, P. Nel, A.A. Prokin, T. Bourgoin, J. Prokop, J. Szwedo, D. Azar, L. Desutter-Grandcolas, T. Wappler, R. Garrouste, D. Coty, D. Huang, M.S. Engel, and A.G. Kirejtshuk. 2013. The earliest known holometabolous insects. *Nature* 503: 257–261.

Parker, A.R. 1998. Colour in Burgess Shale animals and the effect of light on evolution in the Cambrian. *Proceedings of the Royal Society Biological Sciences* 265: 967–972.

*Raup, D.M. 1991. *Extinction. Bad genes or bad luck?* New York: WW Norton & Company.

Schneider, J.W., and R. Werneburg. 1998. Arthropleura und Diplopoda (Arthropoda) aus dem Unter-Rotliegend (Unter-Perm, Assel) des Thüringer Waldes (Südwest-Saale-Senke). *Veröffentlichungen Naturhistorischen Muesum Scheleusingen* 13, 43–53.

*Selosse, M.A. 2017. *Jamais seul—Ces microbes qui construisent les plantes, les animaux et les civilisations*. Arles: Actes Sud.

Song, H., P.B. Wignall, J. Tong, and H. Yin. 2013. Two pulses of extinction during the Permian-Triassic crisis. *Nature Geoscience* 6: 52–56.

Sour-Tovar, F., J.W. Hagadorn, and T. Huitrón-Rubio. 2007. Ediacaran and Cambrian index fossils from Sonora, Mexico. *Palaeontology* 50 (1): 169–175.

*Steyer, S. 2009. *La terre avant les dinosaures*. Paris: Belin.

Stworzewicz, E., J. Szulc, and B.M. Pokryszko. 2009. Late Paleozoic continental gastropods from Poland: Systematic, evolutionary and paleoecological approach. *Journal of Paleontology* 83: 938–945.

Taylor, T.N., H. Hass, W. Remy, and H. Kerp. 1995. The oldest fossil lichen. *Nature* 378: 244.

Taylor, T.N., H. Hass, and H. Kerp. 1999. The oldest fossil ascomycetes. *Nature* 399: 648.

Taylor, T.N., E.L. Taylor, and M. Krings. 2009. *Paleobotany: The biology and evolution of fossil plants*, 2nd ed. Amsterdam: Academic Press.

Terry, I. 2002. Thrips: The primeval pollinators? In *Proceedings of the 7th International Symposium on Thysanoptera*, ed. R. Marullo and L.A. Mound, 157–162. Canberra: Australian National Insect Collection.

Ubaghs, G. 1967. General characters of Echinodermata. In *Treatise of on invertebrate paleontology, Echinodermata 1*, ed. R.C. Moore, 3–60. Geological Society of America and University of Kansas Press.

Valentine, J.W., D. Jablonski, and D.H. Erwin. 1999. Fossils, molecules and embryos: New perspectives on the Cambrian explosion. *Development* 128: 851–859.

Van Roy, P., and D.E.G. Briggs. 2011. A giant Ordovician anomalocaridid. *Nature* 473: 510–513.

*Wohlleben, P. 2016. *The hidden life of trees—A visual celebration of a magnificent world*. Vancouver (Canada): Greystone Books.

Zamora, S., B. Lefebvre, J.J. Álvaro, S. Clausen, O. Elicki, O. Fatka, P. Jell, A. Kouchinsky, J.-P. Lin, E. Nardin, R. Parsley, S.V. Rozhnov, J. Sprinkle, C.D. Sumrall, D. Vizcaïno, and A.B. Smith. 2013. Cambrian echinoderm diversity and palaeobiogeography. *Geological Society of London* 38: 157–171.

5

The Mesozoic, The Age of Dinosaurs

5.1 And Now… The Dinosaurs Come In!

The Palaeozoic gave way to a new era: the Mesozoic. Unlike the previous era, the Mesozoic comprises only three periods: the Triassic (~252–201 My), the Jurassic (~201–145 My), and the Cretaceous (~145–66 My). The Mesozoic is also known as the Secondary Era, or simply the Secondary. However, its most famous designation is undoubtedly the one that commemorates the best-known extinct animals of the entire Phanerozoic. In popular imagination, this time interval has become the "Age of the Dinosaurs".[1]

These large reptiles are the prehistoric animals that have fascinated us the most. Who hasn't daydreamed about the imposing bulk of *Tyrannosaurus rex*? This chapter is partially devoted to them. Yet only partially; it must not be forgotten that many other remarkable creatures appeared alongside the dinosaurs. Ignoring them would be as serious as overlooking the large reptiles themselves, given their importance for current biodiversity. Although dinosaurs occupy the dominant place in all treatises on the Mesozoic, likely due to their popularity, it must be emphasized that they were not alone.

Let us proceed in order, beginning with the dinosaurs.

[1] I would like to point out that, in this volume, the term "dinosaurs" refers to those who are called "non avian dinosaurs" by specialists, that is the animals commonly indicated under this label. However, the clarification is now necessary, because we now know that modern birds are only a particular branch of the great reptiles: technically, they are dinosaurs too!

A. M. F. Valli, *The Three Domains of Life*, Copernicus Books,
https://doi.org/10.1007/978-3-032-14802-5_5

This work is not intended to address the subject exhaustively; doing so would detract from the excellent publications already available.[2] I aim only to present a few concepts that I consider important about these animals.

When did dinosaurs appear? The oldest fossils attributed to dinosaurs date back to approximately 230 My (in the early part of the Late Triassic; Allain 2012). Their remains were first discovered in South America. At that time, the continents were still united in Pangaea, allowing dinosaurs to spread and colonize other lands. Throughout the Mesozoic, they appear on every continent: the Americas, Africa, China, Thailand, Australia, Madagascar, passing through India. Of course, dinosaur fossils (not only bone remains but also egg remains) have also been found in Europe, especially in France.[3]

Initially, dinosaurs shared their habitats with many other animals, especially reptiles. The earliest species were not yet the spectacular giants we admire in museums. The first dinosaurs were relatively modest in size, though *Herrerasaurus ischigualastensis*, one of the oldest known, measured about 5 m. They were often preyed upon by more powerful reptilian predators.

How can we recognize dinosaurs among other contemporary reptiles? One defining anatomical feature is their upright posture (Benton 2024). The femur and hip joint of dinosaurs were arranged so that the hind limbs were positioned straight under the body.[4] This bipedal stance may have contributed to the later success of dinosaurs during the Mesozoic. In fact, after the Triassic crisis, dinosaurs, both carnivorous and herbivorous (the latter reaching lengths over eight meters), became the dominant terrestrial vertebrates, following the extinction of many competitors. Luck or evolutionary superiority?

After the crisis, dinosaurs began to diversify spectacularly, throughout their existence. From the Jurassic onwards, the classic forms familiar to us were established: massive herbivorous sauropods with immense necks and

---

[2] The literature devoted to the great Mesozoic reptiles is very extensive, a sign of the interest they have been able to arouse not only among specialists but also in the hearts of ordinary people. I will give you a simple list of works, separating those for professionals from those made for a wider audience. The first three include: Glut (1997), Sereno (1999), Weishampel et al. (2007). The general public can consult: Czerkas and Czerkas (1990), Norman (2000), Buffetaut (2005), Allain (2012). As far as iconography is concerned, I would like to point out the book by Dixon (2006) which presents artistic reconstructions of large reptiles (not only dinosaurs). Finally, for the younger (under 10 years), Romain Amiot has produced a small volume (Amiot 2005) which is sold with a CD-ROM.

[3] When we think of dinosaurs, we always think of the spectacular remains found in North America or China. However, several remains have also been exhumed in France. In this regard, I point out the book by Buffetaut (1995), dedicated to the French specimens. It is a reading for specialists and uninitiated.

[4] All the first dinosaurs were bipedal creatures, agile and slender, able to run erect on their hind legs. Only later did they become quadrupeds, when the vegetarian forms increased in size and had to rest on their forelimbs because of their weight.

tails (*Diplodocus*, *Apatosaurus* [= the brontosaurus or "thundering lizard"], *Mamenchisaurus*, *Amargasaurus*; Fig. 5.1), or taller than a four-storey building (*Brachiosaurus*), and bipedal carnivores with massive heads and sharp teeth (*Allosaurus*, *Megalosaurus*, *Cryolophosaurus*; Fig. 5.2). These species belong to one of the two major dinosaur groups into which dinosaurs are classified.[5] The second one diversified further during the Cretaceous, producing horned, duck-billed species,[6] some resembling rhinoceros with large horns (Fig. 5.1). In total, over 870 species are known, spanning more than 160 My; more than twice the time separating us from their extinction!

A long-standing question about dinosaurs is whether they were "warm-blooded," like mammals and birds, or "cold-blooded," like modern reptiles. Were they capable of producing internal heat, or did their temperature depend on the environment?

The American palaeontologist Robert Bakker (1988) argued that dinosaurs were endothermic, able to generate their own body heat and maintain a high internal temperature regardless of external conditions. Endotherms contrast with ectotherms, whose body temperature varies with their environment.

However, the answer is not simple. Dinosaurs varied greatly in size and lifestyle: from the enormous brachiosaur, weighing 30–50 tons (equal to that of a dozen elephants; see, in Appendix, the Sect. "Large and small dinosaurs"), to the small *Velociraptor*, an agile and slender carnivore (Fig. 5.2). It must be borne in mind, in fact, that each dinosaur could have different needs, each linked to its own activities. The higher the internal temperature of the body, the more active its metabolism is (understood as the set of biological processes within the body that keep it alive). So, what is important is that the metabolism works in such a way as to allow each animal (= dinosaur) to perform all the activities necessary for its survival. The activities of a *Velociraptor*, which has to land prey thanks to a claw placed on the inner finger of

[5] These two groups form the orders of saurischians and ornithischians. The difference lies, among other things, in the morphology of the pelvis. In saurischians (which literally means "pelvis from saurus") the head of the pubis is directed towards the front. In ornithischians (whose name means "bird pelvis"), on the other hand, the head of the pubis is parallel to the ischium (although various forms developed a pre-pube, directed anteriorly). This division was established very early in the history of palaeontology; it is due to the Englishman Harry Govier Seeley, who established it in 1887 (Allain 2012). Despite the antiquity is still used (but see also Baron et al. 2017). Ironically, birds do not come from ornithischian dinosaurs: their ancestors are to be found in the saurischian group!

[6] One of the characteristics of this group of dinosaurs is that they could actually "chew" their plant foods. In fact, they had an anatomical device of the skull that allowed them a kind of chewing. A particular articulation at the level of the upper jaws, allowed these to separate a little during the closing of the mouth. Then, the opposing upper and lower teeth, whose surface was oblique, could slide over each other, producing a friction that helped to crush the vegetables in the oral cavity.

Fig. 5.1 Reconstructions of herbivorous dinosaurs: top left, the hadrosaur *Charonosaurus jiayinensis*, from the Chinese terminal Cretaceous (total length, 13 m); top right, the nodosaurid *Edmontonia longiceps*, which lived in North America, during the late Cetacean (species of the genus *Edmontonia* could measure up to 7 m in length); bottom left, the ceratopsid *Eionosaurus procurvicornis*, of the upper Cetacean of Montana (United States; 6 m of total length); lower right, the sauropod *Amargasaurus cazaui*, of the Argentine Lower Cetacean (12 m long)

Fig. 5.2 Reconstructions of carnivorous dinosaurs: in the upper left, *Cryolophosaurus ellioti*, whose remains have been exhumed in the lower Jurassic sediments of the Antarctic (total length, 6 m); in the upper right, the famous *Velociraptor mongoliensis*, from the Late Cretaceous of China, Mongolia and Russia, represented with feathers, like the other forms of the family (this small carnivore reached two metres in length); at the bottom in the middle, the strange *Masiakasaurus knopfleri*, from Madagascar, who lived at the end of the Cretaceous (length: about 1.8 m)

its hind legs, will certainly be different from those of a large sauropod that peacefully grazes on plants!

It is clear that living in a warm environment, without particular thermal variations, could be convenient for animals the size of a brachiosaurus or a diplodocus, even if their physiology did not allow them to regulate their internal temperature, independent of that of the external environment. Beasts of this size did not need to be "warm-blooded". Their thermal inertia would have been enough to withstand the small thermal variations between day and night.

On the opposite side, the activities of raptors (*Velociraptor*, *Deinonychus*) needed more energy. Their hunting methods, in terms of energy expenditure, were similar to those of a bird of prey or other active predator; they had to find the prey, catch it and subjugate it. The physiology of these dinosaurs must therefore have resembled that of the animals mentioned above.

However, the possession of a high metabolism comes at a cost: more frequent food and the ability to insulate the body to avoid losing the heat produced.

The histological analyses performed on the fossil remains allow us to shed some light on the subject. Studies of archosaurs[7] from the Late Triassic to the Early Jurassic showed that these animals had rapid growth consistent with a high metabolism. But there is one exception: crocodiles. These reptiles, unlike other archosaurs, have preserved an ectodermal metabolism (Allain 2012, Chap. 4). Some dinosaurs may have improved the physiological characteristics of their ancestors, while others may have followed an opposite evolutionary path.

We have evoked the ability to equip oneself with an insulating covering to avoid dispersing body heat. For modern vertebrates, this solution translates into a covering of hair or a layer of fat (as cetaceans and pinnipeds do) in the case of mammals, or a covering of feathers in the case of birds. The latter, thanks to the asymmetrical shape characteristic of their feathers, can also fly.[8]

[7] Archosauria ("dominant reptiles") are a group that includes, among others, dinosaurs, flying reptiles and crocodiles. Among the particular characters that define the group, we note the presence of an ante-orbital window (Benton 2024). Their skull, on each side, has an additional opening between the orbit and the nostril. This orifice, however, may close secondarily in some evolutionary lines (such as that of the crocodiles) or even merge with the orbital window (as in the case of birds).

[8] Birds' feathers fall into two broad categories: the inner feathers and the outer lining feathers. The former are mainly important for thermal insulation, the latter are essential for flight and are asymmetrical. The central axis of the feathers is formed by the rachis, a tube of keratin. On its sides, to the right and left, are inserted the barbs (arranged a bit like the branches of a tree, but much more tightened), which in turn branch out into barbels. The beards on one side are developed differently than those on the other; this arrangement allows the feathers to form a supporting surface and allow flight (Buffetaut 2005). The feathers of birds that have permanently lost their ability to fly (ostriches and other ratites, penguins) have become symmetrical.

And the dinosaurs? What type of insulating cover could they have? Fossils of dinosaurs that have preserved traces of skin are not frequent, but we know of some. Without going into details, let us limit ourselves to the case of thermal insulation. Today we know, thanks to the discoveries made in recent years (such as the exceptional finds in Chinese sediments of the Middle Jurassic/Early Cretaceous age) that some categories of dinosaurs possessed different types of "cloaks": some had a rudimentary covering consisting of simple filaments, some a kind of more elaborate "down" and those who finally, could boast real feathers, although, in general, they lack the asymmetry typical of those of birds.[9] These structures have evolved, over the years, starting from the scales of reptiles. Although in the past they were considered the prerogative of birds, palaeontologists discover, today, that the feather had already been "invented" by dinosaurs.

Advanced anatomical analyses have made it possible to highlight, in some dinosaurs, a complex anatomy at the level of the vertebrae, previously unknown. The specialists noticed that the vertebral body had deep cavities. This is a well-known peculiarity in birds, where it is associated with the presence of respiratory tissues, the air sacs. These allow you to increase the amount of oxygen-rich air that your pet can store in your body. This device allows the bird to "enrich" its blood with oxygen, a precious element for a flying organism that expends a large amount of energy. This surplus can be particularly important when the bird flies at dizzying heights, where oxygen is rarer.

Could that advantage bring dinosaurs this anatomical arrangement, they who must never reach such heights? A font does not always have to confer an advantage on its owner. It is simply enough that it does not compromise its efficiency in the living environment. On the other hand, the presence of air sacs in the vertebrae of dinosaurs could enrich the oxygen content in their blood, just like birds. This ability could prove advantageous for those species requiring a high metabolism, due to their activities.

Here we discover that dinosaurs (at least, some of them) possessed characteristics typical of modern birds. This detail is important for the ability of these animals to produce energy (an attitude intimately linked to the ability to fly) and to retain body heat.

There is nothing strange, therefore, if a dinosaur with all these characteristics at some point took a further step and started flying (Buffetaut 2005, p. 57–59). Here's how birds came onto the scene during the Mesozoic!

---

[9] Among the scientific articles dealing with this topic, I would point out the following: Ji et al. (1998), Xu et al. (2004), Xu et al. (2010). The interesting reader may also wish to consult Chang (2003).

But when? And where? The oldest known bird, *Archaeopteryx lithographica*, come from the Upper Jurassic (~150 My) strata of the German resort of Solnhofen, Bavaria. The animal, of which various complete skeletons are known, possessed a long reptilian tail, jaws with teeth and clawed fingers on the forelimbs. It would have been classified as dinosaurs if the edges of the fossil had not had clear feather prints along the animal's edge.[10] This time (the discovery of the first fossil was in 1860), which said "feathers" said "birds" and, despite the typical characteristics of a reptile, the specimen was classified as a bird. This determination has hardly ever been questioned, and *Archaeopteryx* continues to be considered the oldest known bird.

However, given the number of exceptional fossils,[11] including several species of Mesozoic birds, from the Chinese province of Liaoning, it is practically certain that, sooner or later, older forms may be discovered.[12]

In all cases, there are two important facts to be considered for our history:

(1) first, birds make their appearance during the Secondary Age;
(2) finally, they are nothing more than particular dinosaurs, covered with feathers, which have acquired the ability to fly.

The group will differ throughout the Mesozoic. Various specimens have been found in Chinese cetacean sediments, but more fragmentary and less spectacular remains have also been exhumed in America (in North as well as South) and in Europe. During their evolution, birds lost their long reptilian tails, as well as their teeth (they acquired a real beak in their place) and clawed fingers on their hands. The K/T crisis with which the Cretaceous (and all the Secondary) ends, affects birds like other taxonomic groups. These, in

[10] In addition, subsequent analyses have shown that the feathers of the outer shell of *Archaeopteryx* were asymmetrical. The animal was thus able to fly. I consider this detail important because, in my opinion, a bird is not more than a dinosaur able to fly (even if some birds have lost second tala ability). However, I would stress that this is a strictly personal definition. Special anatomical features, especially cranial ones, are used to differentiate birds and dinosaurs. However, as far as I know, no dinosaurs that can actually fly have been discovered yet.

[11] Among these exceptional fossils, a specimen of *Anchiornis huxleyi*, a feathery dinosaur from the Jurassic, on which palaeontologists have recently found traces of fossil pigments. The colour of the plumage has therefore been partially reconstructed (Li et al. 2010).

[12] Recently, a scientific communication has proclaimed the discovery of possible bird, *Aurornis xui*, even older than *Archaeopteryx* (Godefroit et al. 2013). However, the antiquity of the finding has been questioned: the layer from which the fossil comes may be younger than 35 My (Balter 2013) than acclaimed. Further analysis of the sediments in which the remains were found is therefore necessary to determine their age. However, no one has yet proven that *Aurornis* could fly. So, until next confirmation, for me *Archaeopteryx* continues to be the oldest bird known!

effect, lose a complete group, that of the enantiornithes,[13] which was the most characteristic of the Secondary.

Despite the crisis, birds survived and evolved into modern species. Since birds are the only descendants of dinosaurs—now numbering around 10,000 species—one could argue that the "Age of the Dinosaurs" continues today.

5.2 Not Just Dinosaurs

As mentioned earlier, dinosaurs were not the only evolutionary innovations of the Mesozoic. Numerous other organisms appeared before, during, or after the emergence of these large reptiles.

Among vertebrates, flying reptiles, the pterosaurs, emerged just a few million years after dinosaurs (Wellnhofer 1991; Unwin 2005; Witton 2013). They were the first vertebrates capable of active flight, dominating the skies long before birds evolved, which, according to current knowledge, appeared in the late Jurassic, between 160 and 145 My.

In the oceans, marine reptiles such as ichthyosaurs, plesiosaurs, and pliosaurs thrived (Callaway and Nicholls 1997; Ellis 2003; Everhart 2005; Paul 2022). Ichthyosaurs were even older than dinosaurs, having appeared at least 10 million years earlier. Alongside them, cephalopod molluscs—including octopuses, squids, and nautiluses—became widespread. Among these, ammonites,[14] with their distinctive external shells, were particularly abundant and became essential tools for biochronology.[15] They descended from an older Palaeozoic group and are their only surviving lineage.

None of these organisms survived past the Cretaceous–Tertiary boundary; all went extinct around 66 My, if not earlier (for instance, ichthyosaurs disappeared by the end of the Early Cretaceous).

---

[13] Enantiornithes differ from modern birds in particular anatomical features, especially at the level of thoracic vertebrae and scapular belt. Also typical was the fusion of the metatarsal bones (those of the feet) that would be carried out from the proximal end (from the leg) to the distal one (towards the feet), the opposite direction from that characteristic of modern birds. This group, known only during the Cretaceous, included many different species, between 40 and 50 (although a revision is necessary, because many specimens are known only in a fragmentary way), with very different sizes and forms. To learn more about the evolution of avian, from the Mesozoic until today, I recommend read in Collective work (2014).

[14] For more general information on this group, see Lebrun (2008a, b). To all those interested in the taxonomy of specimens from the Lyon region, I would like to point out the work of Rulleau (2006). The same author has published several volumes on ammonites at *Dédale Éditions*. Finally, I recommend the book by LoMedico Marriott (2024).

[15] Biochronology is the discipline that allows us to attribute a relative age (who is older than who) to ancient rocks and sediments in function of the fossils they include inside them.

However, the Mesozoic was also the period when the first representatives of many groups that remain vital to biodiversity today emerged. Modern corals first appeared during the Middle Triassic, gradually replacing the Palaeozoic forms that disappeared during the Permian crisis. At the same time, various groups of foraminifera arose,[16] most of which survive today.

Modern ostracods also seem to have evolved during the Secondary. These animals are small crustaceans, rarely exceeding a few centimetres, encased in a bivalve carapace composed of two articulated, symmetrical elements. Though seemingly minor, they are highly significant in paleoecology. Many post-Palaeozoic forms are closely linked to specific environmental parameters, such as water depth, salinity, and light conditions, similarly to their modern relatives. Identifying an ostracod species allows scientists to infer its lifestyle and habitat, helping reconstruct ancient aquatic environments and their associated communities. Both marine and freshwater ostracods are known, allowing reconstructions across a range of ecosystems.

Let us stay in the seas. Let us now turn to more familiar animals; sharks. Modern groups, whose oldest representatives may have appeared as early as the end of the Palaeozoic, evolve and differentiate throughout the Mesozoic (Maisey 2000; Cuny and Bénéteau 2013; Long 2025). Most modern shark families and rays were already present at the end of the era, with representatives who, in shape and lifestyle, resemble current species.

Bony fish also diversified, with many modern groups appearing for the first time during the Mesozoic.

Lebanon, during a large part of the Cretaceous, was almost completely submerged. It was located on a continental shelf of the ancient Tethys Ocean, a vast expanse of sea between the North Atlantic (which was still opening) and Japan. Lebanon is a veritable source of information on the evolution of fish faunas (fish fauna), between 95 and 80 My, because exceptional deposits have been found on its territory both for the quality and quantity of the fossils found (Gayet et al. 2012). An unsuspected variety of bony fish, as well as other organisms, has been exhumed from a handful of localities. From coelacanths to eels (the oldest known), from the first swordfish to the oldest tetraodontiformes, the Lebanese deposits seem inexhaustible as far as the novelty of fossils discovered is concerned. A good number of modern families are reported for the first time, right at these sites.

If we turn to the west, we discover another expanse of salt water, which has now disappeared. This is the Inland Sea that cut longitudinally through North America, from the northern part of Canada to the Gulf of Mexico.

[16] Do you remember foraminifera? Otherwise, I recommend that you review footnote 55 of Chap. 4.

Here too the fossil discoveries have been exceptional: marine reptiles, sharks and invertebrates. In addition, many remains of bony fish, all lived during the Late Cretaceous. As in Lebanon, modern families lived alongside forms that have since become completely extinct (Everhart 2005).

We end our marine excursion with a taxonomic group unjustly neglected in popular literature; that of gastropod molluscs, which includes snails and land snails, plus countless other marine species. Within the gastropods, the Cænogastropoda group[17] alone includes 60% of the known species, including murex, cones, porcelain, whelks. While the first representatives appeared during the Palaeozoic, their real diversification began in the Jurassic, with many modern families emerging. Many were formidable predators, feeding on molluscs and echinoderms. Among them, the giant triton (*Charonia tritonis*) could prey the "crown of thorns" starfish (*Acanthasters planci*), a major threat of modern coral reefs. This group, therefore, plays a crucial role in marine biodiversity, both in species richness and ecological impact, with its foundations laid in the Mesozoic.

Now let us move on to the land, and explore the continents of the time. The skies of the early Mesozoic were dominated by insects and pterosaurs—no birds yet, though they would appear before the Cretaceous, derived from dinosaurs. Large reptiles were the most conspicuous terrestrial animals, but they were not alone. In their shadow, more familiar creatures began to appear.

With a bit of luck, you can catch a glimpse of the first batrachians, including some without an apparent tail, which have the shape and size of toads (Rage and Rocek 1989; Milner 1994; Evans and Borsuk-Białynica 2009). They appear during the Upper Triassic and begin to evolve towards contemporary forms. During the Cretaceous we can already recognize modern families. And it is precisely in this period that we also find the largest frog known.[18]

Batrachians were not the only Mesozoic novelties. The first mammals also appeared.

Although mammals are often contrasted with dinosaurs—dinosaurs being the rulers of the Mesozoic and mammals taking over in the Cenozoic—they are nearly as old, emerging a few million years after dinosaurs in the Late Triassic (for the characteristics with which palaeontologists recognize mammals, see Sect. "Mammals and mammaliaforms", in Appendix). Mammals can claim a very long evolutionary history, having coexisted with dinosaurs during the Mesozoic and other organisms during the following era.

[17] About the evolution of the molluscs Cænogastropoda and their diversity, see Ponder et al. (2008).

[18] *Beelzebufo ampinga* is the largest known frog. Its remains have been found in Madagascar, in sediments of the Late Cretaceous (Evans et al. 2006).

Fossil evidence shows that Mesozoic mammals were highly diverse, contrary to past assumptions that they were simple insectivores. They occupied multiple ecological niches. Some had underground habits, such as *Fruitafossor windscheffeli* (a small animal from the Upper Jurassic of North America), others were skilled swimmers, just like beavers. *Castorocauda lutrasimilis* was a Chinese docodont[19] about 40 cm long, which, just like the beaver, had a flattened tail for easier swimming (the reconstitution of the two fossil animals mentioned is presented in Fig. 5.3). Still others lived in trees (a contemporary species of *Castorocauda*, was even able to glide between branches using a skin membrane [patagium] stretched between the limbs and the tail; Fig. 5.4) and there were also those capable of "gnawing" food. Finally, we also found the remains of a mammal the size of a badger (*Repenomamus robustus*; Fig. 5.3), a real carnivore, which had fed on a newborn dinosaur. Was the young reptile still alive when the predator fed on it or was it devoured when it was already a corpse? We don't know, but this is the only known case, so far, of a dinosaur being eaten by a Mesozoic mammal.

From the picture presented above it is clear that the mammals of the Secondary Age were very diverse.[20] Their only limitation was size: the largest did not exceed that of a badger. Nevertheless, mammals were by no means rare during the Mesozoic, especially toward the end of the era, in the Late Cretaceous. In North America, for example, teeth and jaw fragments of

Fig. 5.3 Reconstructions of Mesozoic "mammals": on the left, *Castorocauda lutrasimilis*, from the Upper Jurassic; in the centre, *Fruitafossor windscheffeli*, the size of a ground squirrel; on the right, *Repenomamus robustus*, from the early Cretaceous (China)

[19] The docodonts, characterized by the morphology of their teeth, are still part of the "mammaliaforms", of which they constitute the most recent species. Their external morphology was probably not very different from that of true mammals.

[20] Regarding the evolution of mammals during the Mesozoic, I suggest referring to Sigogneau-Russell (1991) and Kemp (2005). For more detailed information, readers may consult consult J.-L. Hartenberger in Buffetaut (2005), Hu et al. (2005), Martin (2006), Ji et al. (2006), Meng et al. (2006).

Fig. 5.4 *Volaticotherium antiquum*, fossil gliding mammal, thanks to a patagium between its limbs (Jurassic, China); the animal's length was about 35 cm

mammals are counted in the thousands. In Mongolia, in sedimentary layers contemporary with those in the United States, fossils are not limited to incomplete remains; complete skulls and/or skeletons have been found. At least 150 specimens of this type have been exhumed to date (Collective work 1992).

Let us now consider reptiles. The oldest turtles (Li et al. 2008) appear slightly younger than the earliest dinosaurs, as their oldest known remains were found in sediments about 210 My old (although new discoveries may reveal even older turtles!). Although they still had teeth, unlike modern forms (which were soon replaced by a "beak" during the Jurassic), they were already covered by a carapace, the most characteristic feature of the group (Fig. 5.5).

As for lizards, reptiles with similar morphology have been known since the Carboniferous. These were not species related to modern lizards. The

Fig. 5.5 *Proganochelys*, one of the oldest known turtles; European upper Triassic, about 1 m long

group to which the "true lizards" belong, the "squamates", appeared later,[21] when the dinosaurs had already reached more than substantial sizes and were completely replacing the archosaurs that still existed. During the Mesozoic, squamates not only diversified, but also evolved some particular biological traits (previously thought to have emerged much later), such as viviparity, which is documented in lizards from the Lower Cretaceous[22]!

Snakes are a subgroup of lizards. They differ in vertebral shape and some skull peculiarities (related to jaw and skull bone mobility). However, the most noticeable feature of these animals is the strong reduction of their limbs and girdles, which eventually disappear completely. They begin to appear in the mid-Cretaceous, around 100 My (Rage and Escuillié 2003; Rage 2005; Apesteguia and Zaher 2006; Longrich et al. 2012). Some fossils from marine environments suggest that snake ancestors may have been aquatic. However, remains of the same age have also been found in continental deposits. Based on current knowledge, the precursor of snakes may have been either a fossorial reptile or an aquatic animal.

I saved a particular group of archosaurian reptiles for last: the crocodiles (Buffetaut 1982, 1983). Like their "cousins," they underwent spectacular diversification during the Mesozoic, appearing roughly at the same time as dinosaurs and mammals.

When people think of crocodiles, they usually imagine large reptiles frequenting rivers, swamps, and water bodies in warm regions. Although three small species exist today,[23] the preceding sentence has already presented the the modern crocodile accurately. During the Mesozoic, however, these animals had much more varied ecologies (and morphologies).

The earliest forms, from the Late Triassic, were generally unremarkable reptiles. By the Jurassic, they began differentiating and deviating from the modern stereotype. For example, saltwater crocodiles appeared during this period, initially resembling the modern form, but later evolving drastically. *Metriorynchus*, from the Upper Jurassic, was fully adapted to aquatic life

[21] The earliest representatives of squamates make their appearance during the Jurassic (see, for example, Evans et al. [2002]). There is also evidence of presumed Triassic squamates in the literature, but their taxonomy and age have been questioned (Evans 2001; Hutchinson et al. 2012). Squamates are Lepidosauria. This group, in addition to the true lizards, also includes the group of rhynchocephalians which, currently, counts only two (*Sphenodon punctatus* et *S. guntheri*), confined to some islets off New Zealand. However, during the Mesozoic, the order was much more flourishing (see, for example, Phillippe et al. [2004, p. 1–11]).

[22] Viviparous animals give birth to their offspring without laying eggs. For information on the viviparous cases of Mesozoic lizards, see Wang and Evans (2011).

[23] The dwarf crocodile (*Osteolaemus tetraspis*), native to Africa, reaches only two metres. In South America, Cuvier's dwarf caiman (*Paleosuchus palpebrosus*) is even smaller and Schneider's dwarf caiman (*Paleosuchus trigonatus*), is slightly larger.

Fig. 5.6 Left, *Metriorhyncus superciliosus*, marine crocodile from the Jurassic of Europe. Right, *Simosuchus clarki*, from the upper Cretaceous of Madagascar, presumed herbivorous. The two animals are not to scale

(Fig. 5.6): it had lost its heavy armour, and its limbs had become fins, perfectly suited to hunting fish, squid, and other aquatic prey.

During the Cretaceous, crocodiles diversified even further. In the latter half of this period, Argentina and Brazil hosted crocodiles of various shapes and sizes. Their diets included piscivorous, carnivorous, omnivorous, and even possibly herbivorous species (Candeiro and Martinelli 2006; Marinho and Carvalho 2009; Fig. 5.6)!

In Africa, during the Cretaceous, the group also included exceptional representatives. During the Aptian-Albian (~120–100 My), Niger was home to a colossal crocodile, *Sarcosuchus imperator*, estimated at over 10 m long (between 11 and 14) and weighing 10 tons. Like modern crocodiles lying in wait in rivers, this giant preyed upon herbivorous dinosaurs, occasionally consuming fish and smaller vertebrates. It was the apex predator of its time.

One of its contemporaries, living in roughly the same area, had a more unusual appearance: a "duck-billed" crocodile (Fig. 5.7). Compared to *Sarcosuchus*, it was a Lilliputian, measuring no more than one metre in length. Its uniqueness comes from the shape of its head and extreme specialization. *Anatosuchus minor* (here is its full name, which was chosen well judging by the anatomy!) had an expanded snout that formed a beak. Its bones housed a highly developed olfactory system. It also had typical carnivore teeth and unusually long claws. Analysis of its skeletal traits allows reconstruction of its likely ecological role: it frequented humid, muddy environments, hunting small soft-bodied vertebrates such as frogs and fish, locating prey with its sense of smell, and capturing them with its claws.

Of another genus, *Araripesuschus*, several species are known, between Africa and South America. It had a very short snout for a crocodile. This character, together with the teeth, make us think of a rather omnivorous diet.

Fig. 5.7 Left, *Anatosuchus minor*, Aptian-Albian (Niger). Right, reconstruction of the head of *Kaprosuchus saharicus* (postcranial remains unknown). The two animals are not to scale

Detailed analysis of its remains, combining comparative anatomy and tomography,[24] showed that the stages of growth, from juvenile to adult forms, were comparable to those of modern crocodiles. However, its limbs must have been proportionately longer in young individuals. When the crocodile reached adulthood, the proportions returned to normal (for a crocodile).

Kaprosuchus saharicus is a little younger than *Sarcosucus* and *Anatosuchus*. Its remains have been found in Cenomanian sediments (~100–94 My), also in Niger. We only know the skull of this animal (Fig. 5.7) but it is enough for us to understand that it is an exceptional specimen. The mouth is bristling with robust and massive teeth. *Kaprosuchus*, whose translation means "crocodile-boar" bears his name well! Judging by the teeth, it must have been a carnivore. However, we do not know its habitat at all, since we have not found anything of its postcranial skeleton. Based on the size of the garment, the total length is estimated at about five metres.[25]

The remains of *Kaprosuchus* were found in the same strata that contained a fauna of carnivorous dinosaurs typical of Africa (*Spinosaurus* sp., *Rugops primus*, *Carcharodontosaurus*) and various herbivores (rebbachisaurids and titanosaurs).

[24] Tomography is an imaging technique that allows the reconstruction of a three-dimensional object from a series of measurements made on "slices" obtained on the object itself. It is widely used in geophysics and also in medicine because it allows to obtain images of the organs without operating on the patient.

[25] For more detailed information on the African Cretaceous crocodiles, see are Sereno and Larsson (2009).

This brief tour of unusual Cretaceous crocodiles should not overshadow the presence of more conventional crocodiles living in river environments, hunting fish and other prey like modern species.

Nevertheless, the Mesozoic was not just the age of dinosaurs. Although dinosaurs were more diverse than previously thought, they were not alone. Batrachians, lizards, crocodiles, and mammals already made up a significant part of biodiversity. Today, compared to the Mesozoic, terrestrial ecosystems have lost dinosaurs and pterosaurs, but the other groups remain largely present, albeit more differentiated.

As mentioned, the evolutionary history of Phanerozoic biodiversity is primarily the story of successive groups of organisms, while the main ecological roles remain largely unchanged.

5.3 The Parasites in the Time of the Dinosaurs

When we evoke interactions between living beings, we immediately think of carnivore/prey (or, possibly, herbivore/plant) relationships: lions hunt antelopes and feed on them, while goats graze on grass. The lion cannot survive without its prey, whereas the latter would live without predators. Similarly, plants are affected by herbivores but can survive without them.

Another type of unidirectional relationship, in which one species lives at the expense of another, occurs in parasitic interactions.

We encountered possible parasites earlier in Sect. 4.3, when discussing the Devonian ecosystem of Rhynie (Scotland). Now, let us skip tens of millions of years and focus on the Mesozoic: what do we know about parasites during the age of dinosaurs? What about the dinosaurs' own parasites?

Modern animals and plants are attacked by numerous pests, including fungi, insects, nematodes,[26] bacteria, viruses, and more. Every living being (even parasites) is host to other invasive species, some of which are highly specialized. Imagine any organism, the first that comes to mind. Even it, like all others, is not immune to attacks from harmful organisms capable of establishing parasitic symbioses[27] of parasitic types. This is also true for pests. Fleas, which infest dogs, are themselves targeted by smaller parasitic organisms, and the chain continues.

[26] Nematodes are worm-like creatures. Some are parasites, others lead an independent life. These organisms are also known as "round worms", due to the rounded shape that the section of their body offers, when cut transversely.

[27] Symbiosis has been defined in footnote 6 of paragraph 3.1.

If this is the rule today, why would it have been different during the Mesozoic? Who were the little creatures that disturbed other organisms? Fleas and lice are among the most common ectoparasites today. Did similar insects exist during the age of dinosaurs?

Modern fleas comprise about 2500 known species that parasitize warm-blooded vertebrates: mammals and birds. By the latter half of the Mesozoic, these groups were already present, along with feathered dinosaurs and pterosaurs, which are also considered warm-blooded. Despite the presence of potential hosts, fleas are rare in the fossil record of this era: only a few specimens from the Middle Jurassic and Early Cretaceous of China, plus isolated finds from Early Cretaceous Russia (Fig. 5.8) and Australia (Huang et al. 2012).

Chinese specimens are particularly important for understanding the evolution of these ectoparasites.[28] All these Mesozoic forms were "disturbers" that infested feathered and haired vertebrates. Their morphology supports this behaviour: body bristles directed toward the rear and forceps on some limb segments are consistent with ectoparasitic habits. Nevertheless, the

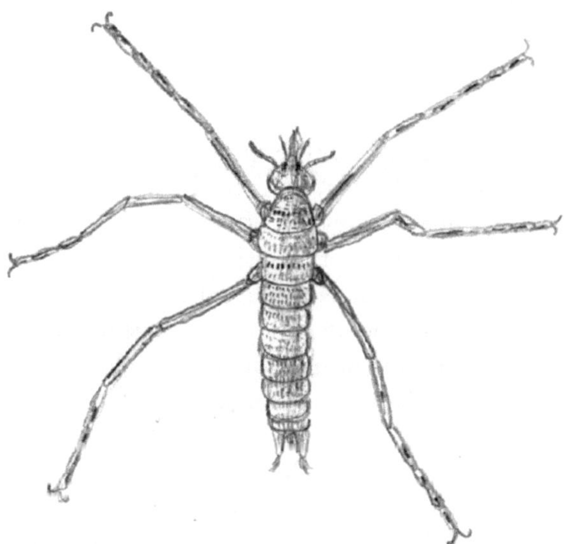

Fig. 5.8 *Saurophthirus longipes*, "flea" from the early Cretaceous of Russia. Probable parasite of pterosaurs and, perhaps, also of the dinosaurs of the time

[28] An ectoparasite is a parasite that remains outside the host's body, like fleas and lice. Endoparasites, on the other hand, infest the inside of the body of their victims, such as intestinal worms, bacteria and viruses.

characteristic anatomical features of fleas and lice must be carefully diagnosed, as misidentifications are common when specimens are few or poorly preserved.[29]

That said, who could be the hosts of such scourges? Their targets were probably the mammals of the time or pterosaurs which, like the former, had a hairy coating (a fur formed of very short hairs). Feathered dinosaurs could also be on the list.

Who were the hosts? Probably the mammals of the time or pterosaurs, which, like the former, had a hairy coating. Feathered dinosaurs may also have been affected. What about the classic, large dinosaurs covered with scales? We can infer some information thanks to "unusual" remains: their fossilized dung (what, in technical terms, is called "coprolite").

The famous Belgian site of Bernissart (Early Cretaceous) has not only provided hundreds of bones of the dinosaur *Iguanodon bernissartensis*, but also fossil droppings, the famous coprolites. The latter have been attributed to the iguanodon, both because of its abundance and because of the size of the droppings. Their analysis made it possible to establish that the dinosaurs were infested with particular nematodes, flukes (another group of parasitic, flat-bodied worms) and protozoa. The latter were present in the form of cysts of characteristic morphology; these organisms were periodically evacuated from the dinosaurs by means of faeces. The work on Belgian specimens has made it possible to describe various parasitic forms; worms and protozoa (Poinar and Boucot 2006). As you can see, the species described by palaeontologists are not always the most spectacular, but also the most unusual!

If we want to continue our review of Mesozoic parasites, we must not neglect another particular source of fossils: amber.

This substance is a resin that is produced by plants, especially conifers. When the plant is injured and its rind is damaged, the deteriorated part exudes this substance in a more or less liquid state. However, it quickly solidifies in the air. During the change of state, he can "capture" an insect or other small creature (or even just a part of it) that will remain, imprisoned forever, inside the resin.

In this way, wonderful specimens of insects, arachnids, and other arthropods have come down to us, perfectly preserved in three dimensions and in all their details, covered with resin that has enclosed them in a kind of magic box, resistant and semi-transparent.

[29] We must be cautious in the assumptions and reconstructions of the alleged fossil parasites. A few years ago, for example, the extraordinary *Strashila incredibilis* was indicated as a possible Mesozoic flea (Grimaldi and Engel 2005, p. 474). However, recently, thanks also to the discovery of new fossil material and the description of a new species, it was discovered that this insect was, in fact, a particular fly, of aquatic habits (Huang et al. 2013).

Remains that do not belong to invertebrates can also be found. In fact, plant fragments are known to be imprisoned in amber, although they are relatively rare. More spectacular are the fossils of vertebrates. An amber specimen from Burma (end of the Early Cretaceous) contains a gecko's foot, while another, found in Canada (Late Cretaceous), contains some feathers. In any case, this type of fossil is extremely rare in amber. Generally, the most common discoveries concern arthropods (insects, spiders, scorpions, centipedes).

Most of the Mesozoic fossil amber deposits that have given us their secrets, the terms of trapped organisms, are of Cretaceous origin.[30] The three most important ones, which will be discussed later in the paragraph, are out of phase in time and space. They offer us three temporal "windows", respectively on three chronological divisions of the Cretaceous: Hauterivian, Albian and Campanian.[31] The oldest deposit is Lebanese, the next is Burmese (Myanmar) while the newest is located in Canada. The resin-producing plants were mainly conifers, especially Araucariaceae and Cheirolepiadaceae,[32] abundant in humid tropical and subtropical regions.

What "secrets" has Cretaceous amber revealed to us[33]? We always go to discover them on the trail of the fossil parasites of the time.

Let us start with those that infest plants. In amber, aphids are well represented. Such insects are known to be vectors of numerous plant viruses, transmitted when animals pierce plant tissues with their mouthparts. Mealybugs are also among the fossil species of Lebanese and Myanmar amber. These

[30] The fact that we speak, in this section, of Cretaceous amber, should not make us forget that there are also other Mesozoic fossiliferous deposits of this resin. I remember the Triassic one discovered near Cortina d'Ampezzo, in the Italian Alps. The site's uniqueness lies not only in its age (~220 My), but also because of the specimens found inside the resin. A whole universe of microorganisms has been revealed to the eyes of specialists: algae, bacteria, protozoa (the latter probably fed on other organisms). A fossil specimen is so similar to the current amoeba, *Centropyxis hirsuta*, that the authors who studied it identified it with the modern species. It is the oldest evidence of a species still existing today. And it is a single-celled organism! (Schmidt et al. 2006). It is also clear evidence that the biodiversity of the Secondary included a rich procession of microorganisms!

[31] Hauterivian, Albian, and Campanian are three chronological divisions (ages) of the Cretaceous. Their respective time spans were: between 133 and 129 My, between 113 and 100 My, and between 84 and 72 My.

[32] The Araucariaceae constituted a family of conifers which, currently, has a distribution limited to a few regions of the southern hemisphere. During the Mesozoic, however, it was widespread worldwide, especially in the Jurassic. The Cheirolepiadaceae, on the other hand, are an exclusively fossil group. They are plants attributed to conifers and made up an important part of the flora between 250 and 70 My. The family groups together different plants: imposing trees and herbaceous plants. Their distinctive feature is the possession of a pollen of a particular type, with specific characters that distinguish it from those of other plants (Taylor et al. 2009, Chap. 22).

[33] Most of the information about the fossils included in Cretaceous amber comes from the book by George O. Poinar junior and Poinar (2008). Later, I will only explain the bibliographical notes concerning other works, able to provide information not present in the volume just cited.

animals are able to damage plants due to toxic secretions. They also carry a good number of pathogens to plants.

The presence of all these insects, which live at the expense of plants, is only an indirect testimony of the possible existence of endoparasites. What is the direct evidence of their presence? Fungi should also be included in the list of plant pests. And, among them, there are also those who have no qualms about parasitizing other fungi. Fossil specimens are not abundant; however, someone has been found! In Myanmar amber, a fossil fungus was found invaded by the hyphae of another fungus, a parasitic species that invaded the tissues of the first (Poinar and Boucot 2007). But the surprises do not stop at this point. The invader, in turn, was attacked by a hyper-parasite (a term used to indicate parasites of parasites), a third species of fungus that attacks the second. All this is perfectly visible (of course if you have the proper techniques) in a drop of Mesozoic amber. In this case, it is really a direct testimony!

Are there others? Amber is truly an exceptional material, which allows the preservation not only of the most minute details, but also of the most delicate structures, such as those of cells and its components. It is thanks to these qualities that, in the trunk of the dipteran *Palaeomyia burmitis*, an insect similar to flies and mosquitoes, imprisoned in Burmese amber, a trypanosome[34] protozoan has been identified. In the bowels of the insect, blood was found. Who was the victim? Modern dipteran of the genus *Sergentomyia* (which belong to the same group of the fossil fly), feed on reptile blood. They are also the vehicle of the *Leishmania* trypanosome. When one of these insects bites a vertebrate to feed on blood, it also sucks up the protozoan (in case its victim is infested). Inside the body of the dipteran, the pathogen transforms to adapt to the new environment, settling in the host's alimentary tract. Finally, during the last stage of the larval cycle, it mutates into a plump cell, equipped with a flagellum, which migrates into the host's head to be transferred to a vertebrate, during the insect's new meal. There, finally, he will be able to complete the last stages of his cycle.

What is interesting is that ALL the known phases of the current protozoan *Leishmania* in the *Sergentomyia* dipteran have been identified in the fossil specimen! During the Cretaceous, some protozoa already possessed complex cycles of existence comparable to those of modern forms.

[34] Trypanosomes are eukaryotic single-celled organisms that can be agents of numerous diseases. The Gambian trypanosome (*Trypanosoma brucei gambiense*), for example, is responsible for sleeping sickness, transmitted to humans by the bite of the tsetse fly (genu *Glossina*).

But the specialists managed to do even better. In another specimen of Burmese amber, they isolated a ceratopogonid insect.[35] Its body cavity was well preserved and provided us with some unsuspected "treasures": corpuscles similar to the cysts of the pathogenic protozoan *Heamoproteus*. Not only that, but inside the wall of the insect's intestine, they identified some tiny objects very similar to polyhedral cytoplasmic virions. These are typical of viruses belonging to particular strains, capable of infecting various types of biting dipterans (hence its presence in the fossil specimen) which, in turn, can transmit them to vertebrates that bite normally.

The direct evidence of a virion is remarkable, given the conditions in which it was found. We just need to find the vertebrate host. What if it was a dinosaur?

5.4 Flowers and Animal Societies

Most of the achievements of the plant world have been produced during the Primary. During the Secondary, there is only one novelty, but of capital importance for the evolution of the group! It is in fact during this era that flowering plants, angiosperms, make their appearance. Today, with 200,000 and 250,000 species, it is the most diverse plant.

When did flowering plants appear, precisely? The oldest definitive traces date to the Early Cretaceous, though some suggest a Jurassic origin, or even earlier.[36]

Among the oldest fossils of flowering plants, *Archaefructus sinensis* (Fig. 5.9) occupies a prominent position, because we know it in its entirety: from the roots to the fronds (Sun et al. 1998; Taylor et al. 2009). Female structures were at branch tips, males below, with small leaves further down. The leaves are small and are located in an even lower position. Although there are no traces of petals or similar structures, *Archaefructus* is considered a true angiosperm. However, its exact taxonomic position within the group is still up for debate. It has been described as an aquatic plant, whose reproductive system must have been underwater.

[35] Ceratopogonid flies are small dipterans that bite, like mosquitoes and horseflies. They are commonly called "biting midges".

[36] The fact that angiosperms, starting from the Early Cretaceous, diversify rapidly, suggests that the group has an even older origin. Some finds of leaves and pollen, dating back to the Triassic or Jurassic, would seem to strengthen this hypothesis (Taylor et al. 2009). However, the first fossil remains considered as true representatives of angiosperms, at least by the majority of palaeobotanists, date back to the Early Cretaceous.

Fig. 5.9 Reconstruction of *Archaefructus siniensis*: the plant measures a few tens of centimetres in all, from the base to the top of the branches. The female reproductive organs are located on the top of the branches and have been represented in a darker colour; the male organs, below the female ones, are clear. The leaves, small, form bunches below the reproduction system

The *Archaefructus* fossils come from a Chinese region that is famous for the discovery of several remains of extinct Mesozoic birds, feathered dinosaurs, and mammals. The set of flora and fauna is known as Jehol Biota. Other plant vestiges also derive from this locality that are included in the group of flowering plants (Chang 2003). Their fossils have been attributed to the genera *Sinocarpus* et *Orchidites*[37] and it is not excluded that future research will not allow others to be found, hidden in the Mesozoic sediments of the region.

[37] However, China does not have the exclusivity of ancient angiosperms. Recently a contemporary fossil of *Archaefructus*, found in the Pyrenees and already known for some time, has been interpreted as an old flowering plant. Its name is *Montsechia vidalii* (Gomez et al. 2016).

Who says “flower”, says “pollination”, especially through particular insects, such as bees and butterflies.[38] But why was this association formed? What advantages do plants gain by using a biological vector rather than another agent, such as water or wind, to carry out fertilization?

To fertilize a plant, pollen, which contains sperm, must pass from the male to female organs. In flowering plants, the male organs are located on the stamens while the females are located in the pistils. On other plants, the sexual organs may be located on cones or other structures. Pollen is carried by an agent that can be biological (a biological being such as an insect or other organism) or abiological (wind, water, or even gravity).

Although it is a more complex problem than the one I am developing here and which must be addressed on a case-by-case basis, it can be said that the method based on the use of a biological vector is much more effective than dispersion by another type of agent. In fact, in the latter case, a large amount of pollen must be produced, most of which is wasted, because it does not reach the target. Instead, the flower only needs a smaller amount. Although it must also produce, in addition, an effective bait, the nectar, the vector then takes care of conducting the pollen to a successful end. Often (but, be careful, it is not always true) pollinating organisms specialize in a limited number of plant species. Thanks to the bait, flowering plants can attract a suitable vector. While the latter feeds on nectar, they attack the pollen on the body. If the vector has been satisfied with the visit to a certain plant, it will look for another of the same species, with the same type of flower. In this way, the pollen will be successful and a new flower will be fertilized.

Thanks to the use of a good dispersing agent and the bait system, the flowering plant is sure that its pollen will arrive at its destination sooner or later. Among other things, the effectiveness of the system is proven by the fact that angiosperms are, nowadays, the dominant plant forms on the planet.

From what we saw in the previous chapter, pollination made thanks to insects would be even older than flowering plants (remember thrips and cycads?). However, angiosperms seem to take over the system and improve it to the maximum. Coincidentally, the oldest fossils of pollinating insects

[38] In reality, the ways of pollinating flowering plants are many and varied. First of all, not all angiosperms reproduce through biological vectors. Some use the wind, just like conifers and other seeded plants (but we must not forget that cycads, although without flowers, use insects to transport their pollen). Among the biological pollinators of angiosperms, we must count, bees and butterflies, of course, various dipterans (this is the order of flies, flies and mosquitoes, which includes various species that feed on pollen) plus other insects (beetles, thrips, and so on). Other known biological agents are birds (hummingbirds in the Americas, nectarinids in Africa, Asia and Australasia, meliphagids in Australasia) and mammals, especially bats and some Australian marsupials (one species, the honey possum, *Tarsipes rostratus* it has reduced dentition and feeds predominantly, not on honey—as the name would suggest—but on nectar and pollen).

par excellence, butterflies and bees,[39] have been found since the Cretaceous (Grimaldi and Engel 2005, Chaps. 11 and 13).

The set of tools set up by angiosperms to attract fertilization vectors prove to be highly effective. Among these, the flower which is a real landing strip, with many types of signals to find nectar (but also many traps and "tricks" to stick pollen on it). It is thanks to all this set of precautions that angiosperms quickly diversify after their appearance.

But we must not forget that, during most of the Mesozoic, the most widespread plants on our planet remain conifers (araucariaceae, pines, sequoias and others; Taylor et al. 2009). These were able to hold their own against competitors for a time, but then they were outclassed by angiosperms. The "marriage" (to use the title of a French book dedicated to the interactions between orchids and pollinating insects[40]) between flower and insect was a success: during the Cenozoic (the era that followed the Secondary), angiosperms became the dominant plants.

The most fascinating example of this marriage is, without a doubt, the relationship that is established between the orchid and one of its peculiar fertilization agents: the bee. When did this interaction start? Is it possible that it has been established since the Mesozoic? Mesozoic bee fossils are not abundant at all. However, we know of some of them. They include, among other things, Cretaceous specimens found in Burmese, Baltic and North American amber.[41]

On the other hand, if we move on to orchids, although the family is one of the most numerous within the angiosperms (more than 18,000 species), their fossils are extremely rare and all post-Mesozoic.[42] Although genetic analysis states that the group had to differentiate before the end of the Cretaceous, no fossils have come to corroborate this hypothesis.

The diversification of flowering plants influenced insect evolution, promoting groups, especially that specialized in pollination. Among these we have mentioned bees. But these insects also have other characteristics than their relationships with angiosperms. In fact, they have another peculiarity that has always fascinated humans, a characteristic that they share with other insects: their social life.

---

39 For information on bees (and only for these insects), see also Collective work (2013).

40 The book by Roguenant et al. (2005) an excellent work for all those seeking information on orchid/pollinating insect relationships.

41 See the footnote 39 of this chapter.

42 The fossil remains of orchids are very rare, both in terms of floral vestiges and pollen. All the finds are later than the Cretaceous. Among these, I mention a Miocene bee, fossilized in Dominican amber, while it was carrying the pollen of an orchid (Ramírez et al. 2007). It is the ultimate proof of the already close relationship between this plant and the bee!

Bees are just one example: ants are also famous for socializing. Just like bees and other insects that specialize in plant pollination, they appear in the first half of the Cretaceous.[43]

What is their relationship with flowering plants? Since ants are negligible as pollination agents,[44] the link is probably indirect. Ants are particular wasps, hyper-specialized in social life. Now, the radiation of hymenopterans (bees, wasps, ants and related forms) is produced as a consequence of the differentiation of angiosperms.

Ultimately, the fact remains that the oldest known traces of social insects were found in the Early Cretaceous,[45] such as those of the first flowering plants.

Ants and bees, because of their social customs, allow me to introduce the last of the great novelties invented by living beings: societies. In the current state of our knowledge, this is an exclusivity of animals. Plants do not organize themselves into society. And neither do the other bodies.

Social life. How can we define it? For a precise definition of this concept, I refer you to specialized texts (for example, the book by Wilson [1975], about sociobiology). However, I emphasize that it implies the fact that various individuals of the same species live together in some particular ways.

If we analyse an animal society more carefully, we can find a system that has already been encountered. Do you remember the theory of the formation of the eukaryotic cell from a symbiotic association of different prokaryotes ($A + A' = B$)? What about the association of various related eukaryotic cells to form a multicellular individual ($B + B + B + \dots + B = C$)? Now, we are witnessing the reunion of individuals of the same species to form a new structure that obeys the following formula: $C + C + \dots + C = D$. D is an animal society, a new entity of a higher order than isolated individuals, obtained, however, by the sum of the latter.

From my scientific readings from my university days, when I was studying in Italy, I learned four types of societies, indicated as follows: colonial invertebrates, social insects, mammalian and bird societies, human societies (Ageno 1986). Let us now see what characterizes them and differentiates them.

[43] The oldest ant fossils date back to the Middle Cretaceous, if not the Lower Cretaceous (Grimaldi and Engel 2005).

[44] *Leporella fimbrata* is an orchid that is fertilized thanks to the ants that visit it (Roguenant et al. 2005).

[45] The most important groups of social insects are those of the hymenopterans, which includes ants (all known species are social), bees and wasps (there are species of bees and wasps that do not live in society) and isopterans, or termites (all social). The latter, starting from the paleontological data available to us, appear during the Early Cretaceous (Grimaldi and Engel 2005). In other groups of insects, there are species that organize themselves into social groups (for example, some thrips). However, we do not know if these are recent or ancient achievements.

A **Colonial invertebrates**. Most of us are familiar with corals, which form the famous coral reefs of tropical seas. These reefs consist of hundreds or even thousands of individual polyps living together in calcium carbonate structures that they build generation after generation. Bryozoans are another type of colonial organism that inhabit exoskeletons they construct themselves.

However, it is not these colonies that I want to focus on here. I want to discuss organisms related to corals and jellyfish: siphonophores, which live in the open sea. One of the most infamous species is *Physalia* (*Physalia physalis*), also known as the "Portuguese man o' war" (Fig. 5.10). It is "infamous" because it possesses powerful venom, which can be delivered simply through contact with its tentacles (siphonophores, like all cnidarians, have stinging cells).

I have used the term "organism," but if we examine *Physalia* closely, we realize it is not a single individual. It is a colony of different beings, all sharing the same genetic heritage. These are many homozygous twins, all derived from the same fertilized cell. Yet, the embryos resulting from the initial divisions follow different developmental paths, just as the different cells of a multicellular organism do. Although genetically identical, these individuals develop distinct morphologies to divide tasks necessary for the

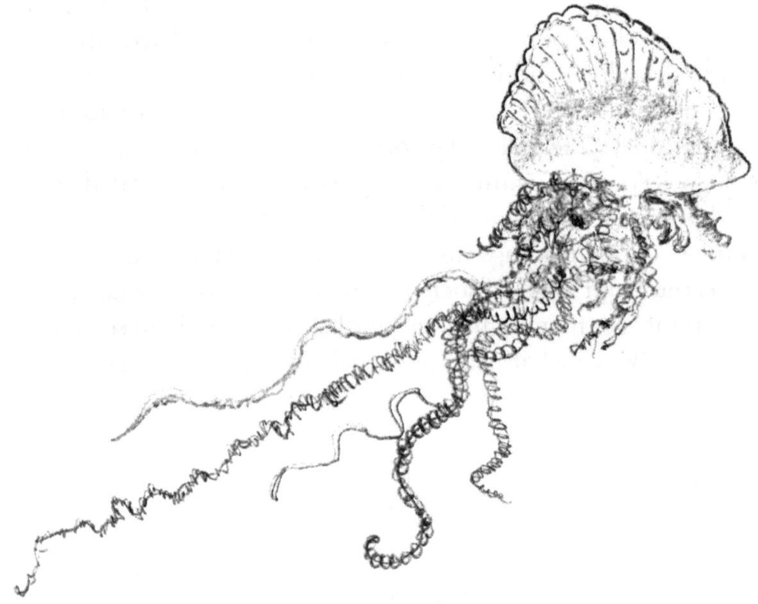

Fig. 5.10 The Portuguese man o' war (*Physalia physalis*) whose tentacles can reach several metres in length

colony's survival: some capture prey, others digest food, and some produce future generations.

Structured this way, *Physalia* is carried by the currents, like all siphonophores, allowing its enormous tentacles (its personal fishing nets) to float like the hair of a sea deity to catch prey.

B **Social insects**. Ants, most bees, various wasps, and termites all form true societies, consisting of hundreds or even thousands of individuals.[46] One of their defining characteristics is a caste system: workers perform daily tasks (searching for food, caring for larvae, cleaning the nest), while soldiers defend the colony. In hymenopterans (bees, wasps, and ants), the worker and soldier castes are female; in termites, they include both sexes.

A queen oversees the colony, her main role being reproduction, assisted by a few males. Again, tasks are divided among different individuals, as in siphonophore colonies. However, here the individuals are independent, and we observe the cohabitation of multiple generations interacting, especially in parental care. For instance, in ants and bees, larvae and nymphs are cared for by their sisters,[47] daughters of the same mother, though the father may differ.

The efficiency of this system is evident from the number and distribution of social insects. Considering ants alone[48] (perhaps the insects most specialized in social life): currently more than 16,000 different living species are known, three times more than that of mammals and at least one and a half times that of birds! Ants inhabit nearly all environments, except mountains above the snowline and polar regions.

C **Bird and mammalian societies**. Many animals live in groups, and parental care typically involves two generations: parents and older offspring. This arrangement serves as an "apprenticeship", allowing individuals to learn how to raise their own young (Cézilly et al. 2006).

Such behaviour is also observed in many bird species and marks the presence of true societies, in contrast to simple colonies, where individuals merely coexist.

[46] African legionary ants of the genus *Dorylus* they can form colonies comprising up to twenty million individuals.

[47] This is important. True animal societies are defined by the fact that more than one generation contributes to the care of the young. In the case of social hymenopterans and termites, the mother is limited to ensuring the birth of the young while the sisters (also the brothers, in the case of termites) take care of the post-birth parental care. This concept of the two generations assisting young people is also essential to the definition of mammal and bird societies.

[48] To learn the secrets of ants, I recommend reading Hölldobler and Wilson (1990). Information regarding the number of known species comes from the Wikipedia site dedicated to ants.

D **Human societies**. The latter are but a special case of mammalian societies. However, it is worth reporting their status and differentiating it from others. We will see why later.

Having reviewed the different associations within the class of animals, let us consider the following question: What are the differences which enable us to establish these four different levels? Of course, some elements have already been pointed out, but the essential difference, in my opinion, lies on another level. It concerns the type of signals that different individuals transmit to each other to maintain the cohesion and effectiveness of society (Ageno 1986).

Let us resume the examination of the four types. The individuals of the siphonophore colonies communicate through tactile (they are physically united) and chemical signals. Thanks to this series of "stimuli" they are all able to ensure the cohesion of the colony and its survival. Social insects continue to use tactile signals (rubbing, food exchange), but chemical signals become more sophisticated (pheromones to recognize the members of the same colony and of different social groups, odorous traces, etc.). Thus, although they are built on the same basis, the signals of social insects become much more complex than those of siphonophores.

In the case of mammalian societies, despite the importance of physical contact (mutual grooming, cuddling) and physical stimulation (marking, personal smells) other forms of communication become essential: vocal and visual signals. In birds, chemical and tactile signals become less important, while visual and vocal communication predominates.

Finally, humans rely on articulated language: it, practically, ensures the totality of information exchanges between individuals and makes interactions increasingly complex and refined. By comparison, all the other "stimuli" take a back seat.

I am perfectly aware of the importance of certain odoriferous or visual signals, especially in particular circumstances (perfumes and clothing to promote seduction; clothes and body care to affirm social status). However, no one can deny that the formation of human societies is based on the communication that derives from our language. It is so sophisticated that it allows us to express abstract concepts. From articulated language emanates culture, or to be more exact, the different types of culture that contribute to differentiating the different human societies.

The possibility of exchanging signals between the different interacting units and the quality of the signals are once again revealed as the determining elements, real milestones capable of introducing real innovations at an evolutionary level. We have already seen this when we dealt with the formation of the eukaryotic cell and on the occasion of the emergence of multicellular organisms. The formation of human societies does nothing but obey the same logic. The result is all the more surprising when the signals are complex and sophisticated: our civilizations are the clearest proof of this.

Before ending the chapter, I have one last clarification. Apart from the colonies of Palaeozoic corals and bryozoans, we have no fossil trace of the first colonies of invertebrates at the level of those of the siphonophores. It is possible that they were formed in the Palaeozoic, or even in the Precambrian. The foundations already existed, especially in the Primary Age. However, evidence is lacking. I took advantage of the discovery of the first social insects to introduce the subject.

And if we move on to vertebrate societies, when do they appear? Were Mesozoic mammals and early birds solitary or colonial? And the dinosaurs?

Although there are indications that large reptiles behaved gregariously (especially for protection or reproduction), and despite the fact that parental care has been shown in some dinosaurs,[49] we do not yet know that they had reached the level of true societies, in which individuals more than one generation take care of the youngest. Taking into account their evolutionary level, we can assume that at least some species were able to do so. But a mere supposition does not constitute proof. Only the progress of scientific research (and new discoveries) will allow us to answer the question.

To those who think that it is impossible to prove the social status of an extinct species, I want to tell a little anecdote. Today's beavers form "family societies" made up of parents and young people of two different generations. Those of the first generation remain in the lair of relatives long after the birth of their brothers (or sisters), in the following year. Well, during the 1990s, in the Oligo-Miocene sediments (~23 My) of the Allier department (France) a burrow with its occupants was exhumed: a family of beavers of the species *Stenofiber eseri*, attested by bones of various individuals (Hugueney and Escuillié 1995, 1996). The study of the fossils[50] made it possible to discover that the remains of the young belonged to individuals of two successive generations, present in the den together with their parents. The result was that this species formed true family societies like modern beavers and that these customs would be maintained for more than 20 million years!

---

[49] The most important indications in this sense come from the study of the footprints that dinosaurs left during their movements or in the places where they laid their eggs (Lockley 1991). However, not all the traces are easy to decipher, especially on the contemporaneousness of all the footprints of the same locality.

[50] However, recently a new discovery appears to be reassembling mammalian societies at the time of the dinosaurs (Weaver et al. 2021).

This shows that even social customs can leave lasting traces in the palaeontological record. Of course, it took an exceptional discovery, but these are less rare than you think: however, you need to know how to recognize and interpret them correctly!

5.5 Summary of This Chapter

The Mesozoic is often referred to as the "age of the dinosaurs" because of the impact that the great reptiles, which lived during this era, had on the collective imagination of both experts and the general public. From their appearance (around 230 million years ago) to their sudden extinction (around 66 million years ago), these fantastic animals evolved spectacularly, producing hundreds of species differing in size and morphology: from giants over 24 m long and weighing several tens of tons, to tiny animals weighing only a few hundred grams and less than a metre long. Some were peaceful herbivores, others fearsome carnivores.

Our knowledge of this group has increased thanks to detailed analyses and fossil discoveries. Exceptional finds have revealed that some dinosaurs were covered with feathers (or proto-feathers), especially on the arms, tail, and even, in some cases, on the hind limbs. Among these forms, we must look for the ancestors of birds, the only "dinosaurs" that survived the great extinction at the end of the Mesozoic unscathed. Birds are the true "heirs" of large reptiles.

But dinosaurs were not alone in their time! Other groups appeared during this era: some survived the extinction at its end, while others suffered the same fate as the dinosaurs (flying reptiles, ichthyosaurs, and other marine reptiles, ammonites, etc.). The descendants of the first groups are part of modern biodiversity. In the seas, we find corals, gastropod molluscs, sharks, and modern bony fish. On land, mammals, turtles, lizards, snakes, and crocodiles thrived. Crocodiles, in particular, were clearly more diverse during the Mesozoic than they are today.

Among plants, the first flowering plants appeared in Cretaceous sediments, the last period of the Mesozoic. Their emergence triggered the evolution of insect groups specialized as pollinators: butterflies, certain dipterans, and bees. Ants also made their first appearance during this period, bringing with them sociality (all modern ants are social: social behavior is intrinsic to their characteristics).

Our knowledge of the Mesozoic world is not limited to vertebrate or insect remains (which, despite popular belief, are far from rare). The exceptional preservation of some specimens (especially those trapped in amber) has allowed specialists to identify parasitic fungi, eukaryotic microorganisms, prokaryotes, and even viruses; all dating back to the Mesozoic. These discoveries have confirmed the existence of previously hypothesized links (especially parasitic relationships) among the different organisms of the era.

Appendix

Large and Small Dinosaurs

With its enormous size (27 m long and 12 m high) the brachiosaurus is undoubtedly the most imposing dinosaur for which a complete skeleton is known. It is also one of the largest known species. However, it is not the largest. For example, *Argentinosaurus*, a sauropod that lived (as its name suggests) in Argentina during the Middle Cretaceous, is estimated to have weighed up to 80 tons and exceeded 35 m in length. Interestingly, the fossil of the largest known dinosaur is not exotic: it is an immense femur (2.2 m long) found near Angeac (Charente, France). Exceptional dinosaurs are sometimes closer than we think! At the other end of the scale, among the smallest dinosaurs, there was *Compsognathus*, a small carnivorous dinosaur about one metre long, weighing between 300 g and 1 kilo. *Compsognathus* lived in Europe, with a specimen discovered in Canjuers, France. Even smaller species have been discovered: *Microraptor*, from the Early Cretaceous of China, is well known for the feathers covering its arms and hind limbs, forming two pairs of "wings" that allowed it to glide between branches. *Microraptor* weighed less than a kilo and measured about 90 cm in length. Even smaller was *Epidexipteryx hui*, slightly older than *Microraptor*, also found in China. This tiny dinosaur weighed only 165 g and measured no more than 25 cm in length, plus about twenty centimetres due to long feather-like structures on its tail. *Epidexipteryx hui* is one of the few arboreal forms known in the group.

Mammals and Mammaliaforms

The class Mammalia has been defined based on characteristics typical of modern species, such as the presence of hair and special parental care (e.g., suckling). However, these features rarely fossilize, except in exceptional cases. Palaeontologists therefore rely on skeletal features. Mammals are the only vertebrates whose jaw joint connects the squamosal bone of the skull with the dentary bone of the mandible. The ancestral reptilian joint, present in mammal ancestors, is located between the quadrate (skull) and the articular (mandible). Initially, the presence of the squamosal/dentary joint was sufficient to define mammals, even when a double jaw joint existed. Today, specialists distinguish "mammaliaforms," which retain both reptilian and mammalian jaw joints, from true mammals, which possess only the squamosal/dentary joint. Mammaliaforms represent the earliest forms, persisting until at least the Late Jurassic. The bones of the reptilian jaw joint do not disappear; they are repurposed as the ossicles of the middle ear. The articular becomes the malleus, and the quadrate becomes the incus, integrated alongside the stapes (Beaumont and Cassier 1994).

Angiosperms

Angiosperms are plants characterized by ovules protected within an ovary (the name "angiosperms" derives from Greek and means "hidden seed", as opposed to "gymnosperms" or "naked seed", because the egg is not closed in the ovary) and by flowers, which house the reproductive organs. Some flowers contain only male or female organs, but many species have both sexes in the same flower. After fertilization (which in angiosperms is double, producing both an embryo and nourishing tissue) the flower develops into a fruit derived from the pistil. The fruit contains one or more seeds and serves to disperse them. For further details, see Taylor et al. (2009, Chap. 22).

References[51]

Ageno, M. 1986. *Le radici della biologia*. Milan: Feltrinelli.

*Allain, R. 2012. *Histoires de dinosaures*. Paris: Perrin.

*Amiot, R. 2005. *Les dinosaures* Fleurus, (with CD-ROM).

Apesteguía, S., and H. Zaher. 2006. A Cretaceous terrestrial snake with robust hindlimbs and a sacrum. *Nature* 440: 1037–1040.

*Bakker, R.T. 1988. *The dinosaur heresies*. London: Penguin.

Balter, M. 2013. Authenticity of China's fabulous fossils gets new scrutiny. *Science* 340: 1153–1154.

[51] Works preceded by an asterisk can be read, more or less easily, by people with a non-professional skilling.

Baron, M.G., D.B. Norman, and P.M. Barrett. 2017. A new hypothesis of dinosaur relationships and early dinosaur evolution. *Nature* 543: 501–506.

Beaumont, A., and P. Cassier. 1994. *Biologie animale—Les Cordés, anatomie comparée des Vertébrés*, 6th ed. Paris: Dunod.

Benton, M.J. 2024. *Vertebrate palaeontology*, 5th ed. Hoboken (New Jersey): Wiley.

Buffetaut, É. 1982. Radiation évolutive, paléoécologie et biogéographie des Crocodiliens Mésosuchiens. *Mémoires De La Société Géologique De France* 142: 1–88.

Buffetaut, É. 1983. L'évolution des Crocodiles. In *Les animaux disparus*, ed. P. Taquet, 102–110. Paris: Belin.

*Buffetaut, É. 1995. *Dinosaures de France*. Paris: Editions du B.R.G.M.

*Buffetaut, É., ed. 2005. *Le monde des dinosaures*. Paris: Dossier pour la Science.

Callaway, J.M., and E.L. Nicholls. 1997. *Ancient marine reptiles*. London: Academic Press.

Candeiro, C.R.A., and A.G. Martinelli. 2006. A review of paleogeographical and chronostratigraphical distribution of mesoeucrocodylian species from the upper Cretaceous beds from the Bauru (Brazil) and Neuquén (Argentina) groups, southern South America. *Journal of South America Earth Sciences* 22: 116–129.

*Cézilly, F., L.-A. Giraldeau, and G. Théraulaz. 2006. *Les sociétés animales: Lions, fourmis et ouistitis*. Paris: Editions Le Pommier/Cité des sciences et de l'industrie.

Chang, M.-M. (ed). 2003. *The Jehol Biota—The emergence of feathered dinosaurs, beaked birds, and flowering plants*. Shanghai Scientifical and Technical Publishers.

*Collective work. 1992. *Dinosaures et Mammifères du Désert du Gobi*, 311. Paris: Muséum National d'Histoire Naturelle.

*Collective work. 2013. *La vie extraordinaire des abeilles*, 175. Sciences et Avenir (July–August).

*Collective work. 2014. *Les ailes de l'évolution*. Espèces, H.S. 1.

*Cuny, J., and A. Bénéteau. 2013. *Requins—De la préhistoire à nos jours*. Paris: Belin.

*Czerkas, S.J., and S.A. Czerkas. 1990. *Dinosaurs—A global view*. New York: Barnes and Noble.

*Dixon, D. 2006. *The Illustrated encyclopedia of dinosaurs*. Leicester: Southwater.

Ellis, R. 2003. *Sea dragons: Predators of the prehistoric oceans*. Lawrence: University Press of Kansas.

Evans, S.E. 2001. The early Triassic Lizard *Colubrifer campi*: A reassessment. *Palaeontology* 44 (5): 1033–1041.

Evans, S.E., and M. Borsuk-Białynicka. 2009. The early Triassic stem-frog Czatkobatrachus from Poland. *Acta Palaeontologica Polonica* 65: 79–105.

Evans, S.E., G.V.A. Prasad, and B. K. Manhas. 2002. Fossil lizards the Jurassic Kota formation of India. *Journal of Vertebrate Paleontology* 22 (2): 299–312.

Evans, S.E., M.E.H. Jones, and D.W. Krause. 2006. A giant frog with South American affinities from the late Cretaceous of Madagascar. *Proceedings of the National Academy of Science USA* 105 (8): 2951–2956.

*Everhart, M.J. 2005. *Oceans of Kansas—A natural history of the western interior sea*. Bloomington and Indianapolis: Indiana University Press.

*Gayet, M., P. Abi Saad, and O. Gaudant. 2012. *Les fossiles du Liban.* 2[a] ed, Désiris.
Glut, D.F. 1997. *Dinosaurs—The encyclopedia.* McFarland & Company, Jefferson (USA) and all its supplements (1–5, between 2000 and 2008).
Godefroit, P., A. Cau, H.D. Yu, F. Escuillié, W. Wenhao, and G. Dyke. 2013. A Jurassic avialan dinosaur from China resolves the early phylogenetic history of birds. *Nature* 498: 359–362.
Gomez, B., V. Daviero-Gomez, D. Coiffard, C. Martín-Closas, and D.L. Dilcher. 2016. Montsechia an ancient aquatic angiosperm. *Proceedings of the National Academy of Sciences USA* 112 (35): 10985–10988.
Grimaldi, D., and M. Engel. 2005. *Evolution of the insects.* Cambridge: Cambridge University Press.
Hölldobler, B.K., and E.O. Wilson. 1990. *The ants.* Cham (Switzerland): Springer Nature.
Hu, Y., J. Meng, Y. Wang, and C. Li. 2005. Large Mesozoic mammals fed on young dinosaurs. *Nature* 433: 149–152.
Huang, D.Y., M.S. Engel, C.Y. Cai, H. Wu, and A. Nel. 2012. Diverse transitional giant fleas from the Mesozoic era of China. *Nature* 483: 201–204.
Huang, D.Y., A. Nel, C.Y. Cai, Q. Lin, and M.S. Engel. 2013. Amphibious flies and paedomorphism in the Jurassic period. *Nature* 495: 94–97.
Hugueney, M., and F. Escuillié. 1995. K-strategy and adaptative specialization in *Stenofiber* from Montaigu-le-Blin (dept. Allier, France; Lower Miocene, MN 2a, ± 23 Ma): First evidence of fossil life-history strategies in castorid rodents. *Palaeogeography, Palaeoclimatology, Palaeecology* 113: 217–225.
Hugueney, M., and F. Escuillié. 1996. Fossil evidence for the origin of behavioral strategies in early Miocene Castroridae, and their role in the evolution of the family. *Paleobiology* 22 (4): 507–513.
Hutchinson, M.N., A. Skinner, and M.S.Y. Lee. 2012. *Tikiguania* and the antiquity of squamate reptiles (lizards and snakes). *Biology Letters* 8 (4): 665–669.
Ji, Q., P.J. Currie, M.A. Norell, and S.-A. Ji. 1998. Two feathered dinosaurs from Northeastern China. *Nature* 393: 753–761.
Ji, Q., Z.-X. Luo, C.-X. Yuan, and A.R. Tabrum. 2006. A swimming mammaliaform from the Middle Jurassic and ecomorphological diversity of early mammals. *Science* 311: 1123–1127.
Kemp, T.S. 2005. *The origin and evolution of Mammals.* Oxford University Press.
*Lebrun, P. 2008a. *Ammonites du Crétacé*, 16. Hors-série: Minéraux and Fossiles.
*Lebrun, P. 2008b. *Ammonites du Jurassique*, 24. Hors-série: Minéraux and Fossiles.
Li, C., X.-C. Wu, O. Rieppel, L.-T. Wang, and L.-J. Zhao. 2008. An ancestral turtle from the late Triassic of southwestern China. *Nature* 456: 497–501.
Li, Q., K.-Q. Gao, J. Vinther, M.D. Shawkey, J.A. Clarke, L. D'Alba, Q. Meng, D.E. G. Briggs, and R.O. Prum. 2010. Plumage color patterns of an extinct dinosaur. *Nature* 127: 1369–1372.
Lockley, M. 1991. *Tracking Dinosaurs. A new look at un ancient world.* Cambridge: Cambridge University Press.

LoMedico Marriott, K. 2024. *Evolution of the ammonoids*. Boca Raton (Florida, USA): CRC Press.

*Long, J.A. 2025. *Secret history of sharks—The rise of the ocean's most fearsome predators*. London: Quercus Publishing.

Longrich, N.R., B.-A.S. Bhullar, and J.A. Gauthier. 2012. A transitional snake from the late Cretaceous period of North America. *Nature* 488: 205–208.

*Maisey, J.G. 2000. *Discovering fossil fishes*. New York: Westview Press.

Marinho, T.S., and I.S. Carvalho. 2009. An armadillo-like sphagesaurid crocodyliform from the late Cretaceous of Brazil. *Journal of South America Earth Sciences* 27: 36–41.

Martin, T. 2006. Early mammalian evolutionary experiments. *Science* 311: 1109–1110.

Meng, J., Y. Hu, Y. Wang, X. Wang, and C. Li. 2006. A Mesozoic gliding mammal from Northeastern China. *Nature* 444: 889–893.

Milner, A.R. 1994. Late Triassic and Jurassic amphibian fossil record. In *In the shadow of Dinosaurs*, ed. N.C. Fraser and H.-D. Suews, 5–22. Cambridge: Cambridge University Press.

*Norman, D. 2000. *Illustrated encyclopedia of dinosaurs*. London: Salamander Book.

Paul, G.S. 2022. *The Princeton field guide to Mesozoic sea reptiles*. Princeton University Press.

*Philippe, M., D. Besson, and D. Berthet. (eds). 2004. *Fossiles de Cerin*. Un, Deux...Quatre, Clermont-Ferrand.

Poinar, G.O., Jr., and A.J. Boucot. 2006. Evidence of intestinal parasites of dinosaurs. *Parasitology* 11: 245–249.

Poinar, G.O., Jr., and R. Boucot. 2007. Evidence of mycoparasitism and hypermycoparasitism in early Cretaceous Amber. *Mycological Research* 111: 503–506.

Poinar, G.O., Jr., and R. Poinar. 2008. *What bugged the Dinosaur? Insects, diseases, and death in the cretaceous*. Princeton University Press.

Ponder, W.F., D. Colgan, J.M. Healy, A. Nützel, L.R.L. Simone, and E. E. Strong. 2008. Cænogastropoda. In *Phylogeny and evolution of the molluscan*, ed. W.F. Ponder and D.R. Lindebrg, 331–383. Berkeley: University of California Press.

Rage, J.-C., and F. Escuillié. 2003. *The Cenomanian: Stage of hindlimbed snakes*, 1–11. Carnets de Géologie/Notebooks on Geology.

Rage, J.-C., and Z. Rocek. 1989. Redescription of *Triadobatrachus massinoti* (Piveteau, 1936) an anuran amphibian from the early Triassic. *Palaeontographica* 206 (1–3): 1–16.

Ramírez, S.R., B. Gravendeel, R.B. Singer, C.R. Marshall, and N.E. Pierce. 2007. Dating the origin of the Orchidaceae from a fossil orchid with its pollinator. *Nature* 448: 1042–1045.

*Roguenant, A., A. Raynal-Roques, and Y. Sell. 2005. *Un amour d'Orchidée: Le mariage de la fleur et de l'insecte*. Paris: Belin.

Rulleau, L. 2006. *Biostratigraphie et paléoécologie du Lias supérieur et du Dogger de la région lyonnaise* Dédale Éditions, Lyon.

Schmidt, A.R., E. Ragazzi, O. Coppellotti, and G. Roghi. 2006. A microworld in Triassic amber. *Nature* 444: 835.

Sereno, P.C. 1999. The evolution of dinosaurs. *Science* 284: 2137–2147.

Sereno, P.C., and H. Larsson. 2009. Cretaceous crocodyliforms from the Sahara. *Zookeys* 28: 1–143.

Sigogneau-Russell, D. 1991. *Les mammifères au temps des dinosaures*. Paris: Masson.

Sun, G., D.L. Dilcher, S. Zheng, and Z. Zhou. 1998. In search of the first flower: A Jurassic angiosperm. *Archaefructus, from Northeast China Science* 282: 1692–1695.

Taylor, T.N., E.L. Taylor, and M. Krings. 2009. *Paleobotany: The biology and evolution of fossil plants*, 2a ed. Amsterdam: Academic Press.

Unwin, D.M. 2005. *The pterosaurs—From deep time*. New York: Pi Press.

Wang, Y., and S.E. Evans. 2011. A gravid lizard from the cretaceous of China and the early history of squamate viviparity. *Naturwissenschaften* 98: 739–743.

Weaver, L.N., D.J. Varricchio, E.J. Sargis, M. Chen, W.J. Chen, and G.P. Wilson Mantilla. 2021. Early mammalian social behaviour revealed by multituberculates from a dinosaur nesting site. *Nature Ecology and Evolution* 5: 32–37.

Weishampel, D.B., P. Dodson, and H. Osmólska. 2007. *The Dinosauria*, 2nd ed. Berkeley: University of California Press.

*Wellnhofer, P. 1991. *The illustrated encyclopedia of pterosaurs*. London: Salamander Book.

Wilson, E.O. 1975. *Sociobiology—The new synthesis*. Cambridge (Massachusetts): Harvard University Press.

Witton, M.P. 2013. *Pterosaurs: Natural history, evolution*. Anatomy: Princeton University Press.

Xu, X., M.A. Norell, X. Kuang, X. Wang, Q. Zhao, and C. Jia. 2004. Basal tyrannosaurus from China and evidence for protofeathers in tyrannosauroids. *Nature* 431: 680–684.

Xu, X., X. Zheng, and H. You. 2010. Exceptional dinosaur fossils show ontogenetic development of early feathers. *Nature* 463: 1338–1341.

6

The Cenozoic, the Era of "Recent Life"

6.1 The Dawn of a New Era

The Cenozoic is the last era of the Phanerozoic Eon, spanning from 66 My ago (the end of the Mesozoic) to the present. According to the International Chronostratigraphic Scale, it includes the following periods: the Palaeogene, the Neogene and the Quaternary.[1]

The world, at the beginning of the Cenozoic, resembled ours much more than that of the Mesozoic and Palaeozoic. The Atlantic Ocean is now open, separating the Americas from Africa and Eurasia. India moved closer to Asia, initiating the collision (since the Palaeogene) that formed the Himalayas and the whole of the southern continents is on the verge of disintegration

[1] Tertiary and Quaternary are no longer eras, but have become divisions of the Cenozoic. The Tertiary is the chronological interval that includes the Palaeogene (between 66 and 23 My) and the Neogene (between 23 and 2.6 My). The Quaternary, on the other hand, is the third Cenozoic period. Recently, a term for a new era, the Anthropocene, has been popularized. Its beginning would be placed at the beginning of the industrial revolution, towards the end of the eighteenth century, to underline the imprint of man on nature from that moment on. However, as far as I know, this proposition has not yet been accepted by the scientific community. The international timescale, for example, does not provide for it. On the other hand, all geologists and palaeontologists who are interested in the Quaternary know well that the "traces" left by man begin to become important as early as the beginning of the Holocene, about 10,000 years ago, well before, therefore, the industrial revolution. Some climatology works tend to attribute anthropogenic "responsibility" to current climate trends starting from very remote periods of human civilizations, between 8000 and 5000 years (Ruddiman 2003). However, it should also be noted that this hypothesis does not make the unanimity of the international scientific community. Regarding the International Chronostratigraphic Scale, see note 22 of Chap. 2.

A. M. F. Valli, *The Three Domains of Life*, Copernicus Books,
https://doi.org/10.1007/978-3-032-14802-5_6

(contacts between Antarctica and South America, on the one hand, and Australia, on the other, become more and more tenuous, until they cease completely). Nevertheless, at least at the beginning, there is still some difference between the geographical framework of the time and the present one. For example, a vast expanse of water, the Tethys Ocean, still stretched between Indonesia and Gibraltar. North and South America remained separated for most of the Cenozoic.

And what about the temperatures? What kind of climate was there on our planet during the Cenozoic? During the first phase of the Paleogene, at the beginning of the era, the average temperature was higher than today. A tropical (or sub-tropical) climate was widespread over most of the landmasses of the time (Zachos et al. 2001; Hansen and Sato 2012). But these conditions will not remain constant forever. Starting from 50 My, we see a drop in average temperatures and, although there will be other rises, temperatures will no longer reach the peaks of the early Palaeogene.

The climatic changes of the Cenozoic, especially during its last period, the Quaternary, are essential to understand the dynamics of the variations that occur in our epoch and to evaluate its future evolutions.

How do we measure the temperatures of ancient times? What techniques allow us to do this? There is no shortage of tools. The most widely used technique is based on oxygen isotope analysis.

Do you remember the definition of isotopes? I already talked about this at the beginning of Chap. 2, where carbon isotopes have been introduced.[2] Isotopes are atoms of a chemical species that differ from each other in the number of neutrons, therefore in their mass. Because of this difference, they are used in a variety of applications.

To assess past temperatures, the two stable isotopes of oxygen, called ^{16}O and ^{18}O, are considered. The first, the most abundant in nature, has as many neutrons as protons, eight and eight, for an atomic mass equal to 16. The second, on the other hand, has eight protons, exactly like the other, but has two additional neutrons: ten in all. So, it's a bit heavier than the previous one. Compared to the first, ^{18}O is relatively rare. Their ratio in nature is known and expressed in parts per thousand. The value is indicated by the symbol $\delta^{18}O$.

When researchers measure the $\delta^{18}O$ contained in a molecule, they evaluate the amount of ^{18}O versus that of ^{16}O. The value can be equal, lower or higher than that of the natural ratio, since the result depends on the reactions that allowed the realization of the molecule under analysis and the quantities of

[2] More specifically, the isotopes of chemical elements have been introduced in the first part of paragraph 2.1.

the two isotopes present in its precursors. The relationships deriving from biological chemical reactions are known, thanks to the analyses and studies carried out in the tissues and molecules of living beings.

For example, the analysis of the coating of a modern foraminifer[3] which lives in the ocean floor, gives us a very precise and known value, under standard conditions. This result corresponds to the biological reaction that consists of integrating oxygen atoms taken from water into its outer coating. The final isotope ratio also depends on the amounts of ^{16}O and ^{18}O present in the liquid environment of the organism. These amounts may vary with temperature.

Let us see how. First, we know that the two isotopes can react slightly differently during chemical reactions. For example, ^{16}O being lighter than the other, evaporates more easily. So, in water vapor, the amount of ^{16}O will be a little more abundant than that of ^{18}O compared to the initial amounts of seawater. As a result, the latter will be a little poorer in ^{16}O. But when it rains, rainwater helps to rebalance the excess of the isotope ^{18}O in ocean waters and restores the normal ratio.

However, during ice ages, or more simply, when it was colder than usual, a greater amount of snow and ice, from the weather, accumulates on the continents. The excess ^{16}O, which was contained in water vapor, is no longer returned to the ocean mass, because it is sequestered in continental snow masses. The ratio between ^{16} and ^{18}O is not more balanced than that of temperate periods: the marine environment, therefore, during cold periods, will exhibit an excess of ^{18}O.

The difference, although minimal, is sufficient to alter the ratio of the two isotopes (in relation to the standard value), during the construction of the lining of the foraminifera built during this period. They too will contain a little more ^{18}O than the standard value. In this case, the $\delta^{18}O$ is slightly higher than the reference value.

On the other hand, during warm periods, snowmelt would be more important and this meteorological water, rich in ^{16}O compared to that of ocean masses, will alter the normal oxygen isotope content of sea water. The constitution of the coating of our foraminifera will in turn be influenced and the value of $\delta^{18}O$, obtained by analysing a sample, will be a little lower than the standard value.

This is how the analysis of foraminifera shells from different eras and the comparison of the results obtained with measurements made on current

[3] For information on foraminifera, see footnote 55 of Chap. 4.

samples allow us to evaluate the increase or decrease in temperature through the various eras, from the past to the present day (Prothero 2006, Chap. 3).

Isotopic measurements can also be made directly from ice water, using similar reasoning. In these cases, hydrogen, the lightest element in the Universe, is used. It is thanks to this type of analysis that we obtained the curves of climate variation during the last glacial phases. The limit of the methodology corresponds to the continuity of the accumulation of snow and ice on the continents, especially in the Arctic and the Antarctic.[4] Beyond that, we must resume analysing the coatings of foraminifera or other marine organisms.

Having made this clarification, let us return to examining biodiversity and analyse the situation at the Cenozoic. In the seas, the large marine reptiles, characteristic of the Mesozoic, have become extinct. Among them, only turtles have been able to overcome the K/T crisis.

Invertebrates have also suffered losses: ammonites, cephalopods typical of the Secondary, are no longer there, as are other groups of molluscs, including rudists.[5] Gastropods, on the other hand, are booming. Bony fish and sharks, despite the losses (the first group may have lost up to 84% of the known species) continue their evolution and diversification in the seas and oceans of the globe.

With regard to corals, the representatives of the current groups (which had already made their first appearance in the first part of the Mesozoic) began, together with other marine organisms, the construction of modern biotopes.[6]

It is precisely during the Cenozoic that coral reefs more or less similar to those we still encounter today in warm waters begin to be built. Modern barriers were formed during the last glaciations (about 10,000 years ago). The fossils, however, tell us a story of even older coral reefs. That of the fish of Monte Bolca (an Italian town near Verona), aged close to 50 My, is considered the oldest which, due to its structure and type of taxonomic composition, represents a modern coral reef fauna (Bellwood 1996).

---

[4] Arctic sea ice would have begun to form 13 million years ago. But the current configuration would not have been acquired until around 3 My (although during the different glacial and interglacial phases its total extension changed). Instead, the Antarctic would have begun to be covered with perennial ice around 15 My. The current cap would have about 6 My.

[5] Rudists are part of the group of bivalve shells (like oysters and clams), but many of them have a particular morphology. The shell that the mollusc uses to attach itself to the substrate (generally the left valve) becomes massive and can have different shapes depending on the family to which it belongs. The other, on the other hand, is usually less developed and serves as an operculum. Some rudists have become true reef-building agents in the Mesozoic seas, just like corals.

[6] In biology, the term "biotope" indicates a living environment defined by its particular physical (such as temperature or humidity) and chemical characteristics.

Coral reef environments are extremely important for biodiversity: they are thought to conserve 25% of today's marine species. In this biome we can observe another close association between different organisms: coral polyps and unicellular zooxanthellae algae.[7] It is thanks to this association that corals have been able to build reefs. Zooxanthellae, in fact, live inside the tissues of polyps. They absorb the CO_2 produced by the animals and also receive their waste, from which they obtain the nitrogen compounds necessary for their development. In return, they provide their guests with different nutrients.

And on the continents? We have seen that, at the beginning of this era, the climate was much warmer than the current one. Vast tropical forests stretched up to the high latitudes. Flora resembled modern plants, though some tropical species were at higher latitudes.[8]

For insects, as for other classes of terrestrial invertebrates, the Secondary had ended without too many losses. Their relationships with plants had strengthened well before the end of the Mesozoic. Insects continued their evolution and continued to weave increasingly close relationships with other organisms, especially plants.

Among vertebrates, saurians and batrachians seem to emerge from the K/T crisis better than birds and mammals. However, the former, despite having lost their most important Mesozoic group, recover quickly.

In any case, dinosaurs have left a void. Crocodiles have also suffered losses. Although there are still numerous species throughout the Cenozoic, they will no longer reach the heights of diversity they had in the Secondary. They specialized in the niche of aquatic predators, with amphibious habits, which they still occupy today.[9]

The Cenozoic is often referred to as the Age of Mammals, because this group of vertebrates is considered to replace the dinosaurs. Appearing at

---

[7] Zooxanthellae are microorganisms capable of photosynthesis (they absorb blue light from the electromagnetic spectrum). Most of them live in symbiotic association with marine animals: mainly corals, but also clams, other molluscs and various species of jellyfish. Corals that build coral reefs (called "hermatypic corals"), especially madrepores, are associated with zooxanthellae. This union is so close that the latter, in some species of corals, are already present in the egg, immediately after fertilization (Selosse 2017, Chaps. 5 and 10).

[8] The determinations made some time ago on the flora of the beginning of the Cenozoic must be considered with caution, because they are mainly based on the morphology of the leaves, a character that is now considered insufficient for correct identification. For further details on the flora of the period, see Taylor et al. (2009). With regard to the French flora, the interested reader can profitably consult the classic volume by Piton (1940) which, however, must be examined with a critical eye because the identifications of plant essences suffer from the problem indicated above.

[9] In reality, things are not that simple. Crocodiles with terrestrial habits are known in South America (the diversity of these reptiles continues to be important, at least on this continent). They are members of the group of sebecids, whose first representatives date from the Secondary era. Terrestrial forms (or considered as such) are also reported in the Paleogene of Europe and North America (the genus *Pristichampsus*) and even Australia.

approximately the same time as large reptiles, they would have remained in the shadow of the dinosaurs, waiting for their moment of glory. In fact, we saw in the previous chapter that mammals had already differentiated throughout the Secondary, both in lifestyles and morphology. In size alone, mammals could not rival dinosaurs.

Like all major groups in the Mesozoic, mammals suffered a number of losses. Only four groups will be able to pass the K/T boundary[10]:

- the marsupials (this group is decimated by the crisis, but we will see that they will show enormous vitality during the Cenozoic);
- the placentals;
- the monotremes (the platypus and echidnas belong to this group, plus some fossil species limited to Australia and South America);
- the multituberculates, a group that will become extinct towards the end of the Eocene between 40 and 30 My.[11]

Despite the losses suffered, mammals continued their evolution throughout the era: during the Cenozoic all modern orders appeared (O'Leary et al. 2013) and, for the first time in their history, these animals began to reach important dimensions, larger than those of a badger.

Their history is becoming better known thanks to the new excavations and laboratory studies that are conducted every day by thousands of researchers from all over the world. No other group of vertebrates can count on such impressive "army" of palaeontologists. The results of these studies fill international journals and a good number of books.[12]

It is often said that before mammals reached large sizes, in the early stages of the Cenozoic, they were "subordinated" to giant-sized birds. *Gastornis* in Europe (Buffetaut 1997) and *Diatryma* (often considered a synonym of

---

[10] In fact, some remains from South America suggest that other groups of Mesozoic mammals were able to overcome the K/T limit. These animals will all disappear during the Tertiary (Wilf et al. 2013).

[11] Multituberculates form an order of mammals that appears during the Secondary, in the Middle Jurassic. They were very diverse and abundant during the Cretaceous, both in North America and Asia. Thanks to the anatomy of the best-known members of the group, specialists attribute the eating habits of rodents to them. And it would be precisely the competition with the latter that would have pushed the multituberculates to extinction, during the Paleogene. However, with more than 120 million years of existence, they represent the group of mammals with the longest lifespan, both among the current and fossil forms (Kemp 2005; Yuan et al. 2013).

[12] Here are some scientific readings concerning the mammals of the Cenozoic. Some works deal with a particular country or group, others with a specific era. I quote them in chronological order of publication: Savage and Long (1986); Guérin and Patou-Mathis (1996); Rössner and Heissing (1999); Turner and Antón (2000, 2004); Agusti and Antón (2002); Janis et al. (2005, 2018); Wang et al. (2008, 2013); Werdelin and Sanders (2010).

the former) in North America would have been the great predators of the time, at the top of the food chain. Should we consider this the last gasp of the dinosaurs? Let us not forget that birds are the legitimate survivors of dinosaurs, the only true heirs of large reptiles.

In reality (as often happens in these cases), things are more complicated. For one thing, giant birds are not limited to the prehistoric world; they still exist today. The African ostrich, with a mass that can exceed 100 kilos (for a male individual) and a height of 2.5 m, is the largest modern bird. Its size exceeds the human average: the ostrich is definitely an imposing bird! Of course, it is completely unable to fly.

Other birds of similar size are emus and cassowaries (Australia and New Guinea) and rheas (South America). Although none of them can be considered a ferocious predator (they are frugivorous-vegetarians, although they do not disdain, from time to time, insects or small vertebrates), no one would dare deny that they are large in size. So, there are giant birds that are our contemporaries and are not necessarily formidable predators. On the other hand, who can claim to know exactly what the diet of the large extinct land birds was?

But let us proceed in order. Let us first consider the presence and distribution of these animals during the Cenozoic. Until relatively recently (about 1000 years ago), there lived in Madagascar a gigantic land bird, whose mass exceeded not only that of the ostrich, but also that of *Gastornis*. It is not without reason that it has been called the "elephant bird" (*Aepyornis maximus*).[13] In New Zealand there lived various moa species (Wilson 2004), land birds of varying sizes, from medium to large (even very large), which filled the ecological niche of large herbivores (as you can see, giant birds are not all carnivores!). One species, the giant moa (*Dinornis giganteus*), despite being lighter than the *Aepyornis*,[14] exceeded it in height, and was also taller than all modern ratites[15]: it is estimated that it could reach three and a half metres! The moa disappeared relatively recently, as did the elephant bird.

What do we know about the evolutionary history of these two groups of giant birds that have almost made it to the modern era? Information about them does not abound. However, in recent years, a Miocene site in New Zealand (presumed to be between 19 and 16 My) has provided bone fragments and eggshell remains that can be attributed to moa (Tennyson et al.

[13] I have already mentioned this animal in Sect. 3.2, when I spoke of the largest cells known. One of these is precisely the egg of this bird!!

[14] The giant moa (*D. giganteus*) weighed ~270 kilograms. The weight of the elephant bird, on the other hand, is estimated at 400 kilos!

[15] With the term "ratites", ornithologists indicate the large land birds (mentioned in the text), plus some other smaller species. They too have lost the ability to fly and share an ancestor with ostriches.

2010). The evolutionary history of these birds can begin to be traced, at least, from the early/middle Miocene.

But the moa and the elephant bird are not the only large land forms known during the Cenozoic. Australia and South America provide us with other examples.

Australia has not only been found the fossil remains of the ancestors of the ratites that live in this country today. The island also provided us with another surprise: it was the cradle of a special group of birds that, in size and mass, could compete with both the moa and the *Aepyornis*. They are called the dromornithids, but they are better known as "thunderbirds"[16] (Fig. 6.1). They are endemic of the Australian continent (they are not known elsewhere) and date back to at least the Late Oligocene (something like 26 My). But while emus and cassowaries are still present among us, the most recent dromornithids date back to the Pleistocene, and would even have encountered the first human colonizers of Australia.[17]

In South America, phorusracids (birds of varying sizes) ranged from around 50 cm to over two metres in height (Alvarenga and Höfling 2003).

Fig. 6.1 Left, reconstruction of *Dromornis stirtoni* (about 2.5 m tall); right, reconstruction of *Phorusrhacos longissimus* (about 1.80 m tall)

[16] An Upper Miocene "thunderbird" from the Northern Territories, *Dromornis stirtoni*, is among the heaviest birds we know: its mass is estimated at about 400 kilos!

[17] On the ancestors of the present-day Australian ratites, see Boles (1997b, 2001). On "thunderbirds", see Vicker-Rich (1979) and Nguyen et al. (2010).

While they are typical of the American continent, some evidence suggests their presence in Europe at the beginning of the Cenozoic (Mourer-Chauviré 1999).

Titanis walleri is one of the most recent species. It lived in Florida about 3 million years ago, although some indications suggest it may have persisted until as recently as 15,000 years ago. *Titanis* was among the animals that migrated to North America from South America after the formation of the Isthmus of Panama. Its size was impressive, reaching up to 2.5 m: comparable to that of an ostrich[18]! The oldest South American species, such as *Paleopsilopterus itaboraiensis* from the Palaeocene of southeastern Brazil, were relatively small. The largest species flourished between approximately 20 and 10 My (Fig. 6.1). With their hooked beaks and claws, phorusracids are universally recognized as carnivorous.

This evidence shows that giant birds were not exclusive to the early Cenozoic. In some regions, they were present throughout much of the era. Elsewhere, their evolutionary history remains poorly understood, though future excavations in Africa or Asia may reveal new insights into the origins of modern ostriches and the extinct elephant bird (Senut and Pickford 1995; Mourer-Chauviré et al. 1996; Bibi et al. 2006).

Not all giant birds were carnivores, however. For instance, moas were herbivorous, and the diets of *Gastornis* (= *Diatryma*) is disputed [19] (were they all really carnivores? Were they omnivores or even herbivores?). Could species with massive beaks feed on carrion or break bones? Like today's ratites, Cenozoic giant birds were integral members of their ecosystems, without necessarily being apex predators, although this may have applied to the larger phorusracids.

What about the Mesozoic? Did giant terrestrial birds exist then? Fossil remains from this era are generally rare and fragmentary. However, a recent study reported a large bird mandible from the Upper Cretaceous of Kazakhstan, suggesting a bird larger than modern ostriches. Named *Samrukia*

[18] See, on the Wikipedia site, at the address dedicated to this bird.

[19] The reader who wishes to learn more about the food nutrition of *Diatryma* and *Gastornis* you can read the following papers: Witmer and Rose (1991); Bourdon and Cracraft (2011); Andors (1992). The first two are fervent supporters of a carnivorous (or simply scavenger) diet for these birds. The last one, however, supports a different thesis. However, recently, an article by Delphine Angst, in collaboration with a Franco-Belgian team (Angst et al. 2014), based on isotopic geochemistry and functional anatomy, brings new evidence about the vegetarian diet of *Gastornis*. It is this the last word on the subject? We will see if these results will be confirmed or if a new study will bring other news! Regarding the feeding of dromornithids, I suggest consulting the website of the "Australian Museum". Finally, I recommend reading the paper by Wroe (1999).

nessovi (the generic name deriving a legendary local phoenix) its classification as a bird is debated. Some argue it could have been a giant pterosaur, as the morphological traits could fit both groups.

So, no giant birds in the Mesozoic? Instead, it seems there was at least one! Late Cretaceous remains from France hint at the existence of *Gargantuavis philoinos*, a large terrestrial bird. While these fossils are incomplete, they indicate that at least some giant birds existed before the Cenozoic. It remains unclear if *Gargantuavis* was related to modern ratites or Cenozoic giants, suggesting it might represent a unique, extinct Mesozoic lineage.[20]

Before I end the paragraph, I would like to make one final point. In this section there has been much discussion about animals. But what do we know about the microorganisms of the Cenozoic (I remind you that bacteria and other unicellular beings make up a substantial part of biodiversity)? Often inconspicuous, their fossil traces are not always easy to identify. Despite this, their vestiges are known even in deposits that have become famous because of the exhumed vertebrates. This is the case of the Middle Eocene site (~47 My) of Messel, Germany. Here, bacterial remains are varied and abundant.[21]

These creatures are often overlooked, due to their small size. But in some cases, their fossils can be very conspicuous. This is the case with the stromatolitic constructions found in limestone quarries containing sediments dating to the Oligo-Miocene of Allier (France); some, cabbage-shaped, can exceed one metre in diameter!

Of course, the best indices to assess their presence are indirect. Animals, especially mammals, harbour a rich community of microorganisms in their bodies, especially in the digestive system. Consider ruminants, which possess, in their stomachs, a rich flora of symbiotic microorganisms essential for the digestion of the cellulose contained in their plant foods.[22] Without the presence of microbes, these animals could not sustain themselves on such a diet.

Our bodies are home to a large amount of these creatures. Each of us hosts more than 1000 different species of microorganisms (bacteria, archaea, protozoa), whose number of cells (estimated, as an order of magnitude, at

---

[20] *Samrukia* has been described in Naish et al. 2012. One of the first works to question its identity as a bird is due to Buffetaut (2011). Finally, the remains of the French Mesozoic giant are described in Buffetaut and Le Loeuff (2010).

[21] Read the section on bacteria, written by K. Liebig, in the work of Von Koenigswald and Storch (1998).

[22] Ruminants have their stomachs divided into four compartments: the rumen, the reticulum, the omasum and the abomasum. The first is also the most voluminous and is the main site of the degradation of plant elements by fermentation, thanks to the microorganisms hosted (Beaumont and Cassier 1994).

a few billion) can vary according to the geographical region inhabited, the age of the individual and other factors. This community helps protect us against certain external pathogens and helps us extract nutrients from food (Lozupone et al. 2012; Selosse 2017, Chaps. 7 and 8).[23]

Now, if we extrapolate this result to the mammalian communities of the Tertiary, we can realize that the diversity of microorganisms of the time must have been no less than that known today.

6.2 Australia's Remarkable Mammal Wildlife

The Cenozoic is often referred to as the mammalian age. I would therefore like to devote this paragraph to the group. However, I do not want to tell you the general history of these animals during the Tertiary and Quaternary. My desire is to limit myself to Australian fossil mammals.

Why choose Australia? Because, currently, it has a fauna of mammals that are unique in the world.[24] With New Guinea (and Tasmania) it is the only place where one can encounter the three mammalian evolutionary lineages that still persist to this day: marsupials, placentals and monotremes. They evolved alongside each other long before humans introduce domestic or wild animals for their own purposes.

Australia is an isolated land. It is the second largest island in the world (after the Antarctica).[25] It has completely separated from the other southern continents, in particular with Antarctica and South America, between 45 and 35 My.

When these masses were still in communication with each other, land animals could move freely from one country to another. But when Australia lost contact, many creatures were unable to cross the water barrier between this land, which had now become an island, and the nearest continent, at the time, the Antarctic. The latter has drifted more and more towards the south and the new climatic conditions that have been established have contributed to making the living conditions of the animals difficult. The specific biodiversity of the Antarctic continent has been greatly reduced. The Australian one,

[23] See also the Wikipedia site dedicated to the microorganisms of the human organism, at the entry "Human microbiota".

[24] For an overview of Australian mammals, native and introduced, plus some species that have become extinct in historical times, I recommend Menkhorst and Knight (2010).

[25] This, of course, is not taken into account by the landmasses. In this case, the blocks that unite, on one side, Eurasia and Africa and, on the other, the two Americas are much larger than both Australia and Antarctica.

on the other hand, evolved independently, in much more clement climatic conditions.

Currently, the closest lands to Australia are those of Southeast Asia.[26] However, although the two continents are actually close to each other, from a geographical point of view, wildlife exchanges are relatively limited. Australian lands are separated from Asia by the "Wallace line", a geographical barrier that passes between Sunda Islands (e.g., Bali, Sumatra, Borneo, and Mindanao) in the northwest and oceanic lands in the southeast. The main obstacles that make it difficult for terrestrial faunas to trade are the oceanic trenches (Asia and Australia belong to different platforms) of the region and the sea currents. This barrier has been working since Australia placed itself in the geographical position it currently occupies.

Australian lands remained isolated during most of the Cenozoic and their mammals were able to evolve separate from the rest of the world. The modern indigenous community is made up of the descendants of the colonizers who arrived in Australia before it broke away from other continents, plus immigrants who were able, from time to time, to overcome the water barriers that separated it from other lands.

The native fauna of the placentals was formed by successive waves, starting with lucky "adventurers" who were able to conquer the island, probably coming from South America and the Antarctic, at first, and from Asia, later.

Modern monotremes, the platypus and echidna species, are the descendants of species that inhabited Australia from the Cenozoic. Their history on the island, alas, still presents many shadows. This could also be because these animals have never been abundant, both in terms of species and individuals. However, we have some fossil evidence that modern families were already present on the island, one, that of the platypus, from at least 28 My, the other, that of the echidnas, in a more recent era (from the Pliocene, about 5 My, or even before; Long et al. 2003; Flannery et al. 2022).[27]

The evolutionary history of Australian marsupials, on the other hand, is better documented. The current representatives may all derive from no more than two populations, who immigrated from the Antarctic before its complete separation. Then they evolved in place (Nilsson et al. 2004,

[26] New Guinea is part of the Australian continent. During glacial periods, when sea levels dropped, the northern lands of Australia and the southern lands of New Guinea could be connected by a land bridge that allowed wildlife exchanges. For all these reasons and for the fact that New Guinea is located on the same side of Australia with respect to Wallace's line (see text), the former is considered as a particular region of the Australian continent, with faunas very similar to those of Australia.

[27] It is necessary to point out that even older monotremes, from the Cretaceous age, inhabited Australia. However, their degree of kinship with the current forms is still poorly understood.

2010).[28] This situation offers palaeontologists a unique opportunity to study the biological evolution and phylogeny of the group in its entirety. In an isolated place, in fact, without migratory flows from other lands, the relationships between ancestors and descendants can be easier to recognize and/or reconstruct.[29]

Until recently, knowledge of Australian marsupials was limited to more recent epochs. Discoveries in northwestern Queensland revealed remarkable faunas from the end of the Oligocene to the Pleistocene, highlighting not only mammals but also birds and other groups (Archer et al. 1994).

Yet, the early Cenozoic remains poorly known. Only a few localities, from the end of the Late Cretaceous (> 66 Ma) to the end of the Palaeocene (~ 55 Ma), provide fossil evidence. For instance, the Murgon site in Queensland may contain remains of the ancestral marsupial lineage and the earliest placental mammals in Australia, including ancient bats and other now-extinct forms.

Some bones found in the Palaeocene locality of Murgon (also in Queensland) may correspond to that of the basic marsupial, from which all Australian forms are derived.[30] Alternatively, the same animal could also be the representative of a group of American opossums and show us that the Australian forms are derived from one of these. Therefore, we are not yet able to correctly classify these Cenozoic fossils in the base of the marsupial phylogenetic tree.

However, the Murgon site is of crucial importance in another respect. It also allows us to document the first population of placental mammals in Australia.[31] Some remains, in fact, would testify to their existence: together with one of the oldest bats known, other bones would prove the presence of unique creatures that became extinct later (Long et al. 2003).

Were there placental mammals that were not flying but endemic in Australia's early history? Here is a fact that, if confirmed, would counter sterile discussions about the alleged superiority of placentals over marsupials.

---

[28] For alternative hypotheses see also Beck (2012).

[29] Australia is not the only continent to have been isolated during the Cenozoic. South America was also in an identical condition, during a large part of the Tertiary, until the formation of the Panamanian ism, around 3 My. This strip of land has re-established contact with the northern continent. The study of fauna and the comparison with that of other countries is another interesting case for all scientific disciplines related to palaeontology and evolution.

[30] The fossil animal community of the locality of Murgon is known as the "local fauna of Tingamarra". For more details on the subject and the marsupials mentioned, see Godthelp et al. (1999) and Beck (2012).

[31] The fauna of Tingamarra is not only crucial to our knowledge of mammals. The remains of the oldest known passerine (Passeriformes) also come from the town of Murgon. This discovery allowed us to shed new light on the origin of the group (Boles 1997a).

A superiority that would be based on two reasons: (1) the current placentals are more numerous and diversified than marsupials; (2) during the faunal exchanges between North and South America, the marsupial species that disappear are proportionally more numerous than those of the placentals.

These reasonings, however, hide the reality. The two groups have equally evolved, each having been able to bring effective and original answers to existential challenges. In fact, in modern Australia, various families of marsupials and placentals coexist without problems: it is rather human interference that endangers the native fauna, more than imported placental mammals.

But now let us turn to the marsupials that once populated Australia and explore their relationships with modern species. Today's marsupial fauna includes many unique animals, such as kangaroos (genera *Macropus*, *Osphranter*), koala (*Phascolarctos cinereus*), wombat (*Vombatus ursinus*), all familiar to the general public and grouped with other herbivorous species in the order Diprotodontia. Less well-known but equally remarkable are bandicoots, curious mammals with pointed snouts and omnivorous diets, classified in a separate order distinct from kangaroos and koalas, and dasyurids, also known as quolls (genus *Dasyurus*), which, along with the Tasmanian devil (*Sarcophilus harrisi*) and the now-extinct thylacine (*Thylacinus cynocephalus*), form the quintessential group of native carnivores. These animals, along with other smaller insectivorous forms, belong to the order Dasyuromorphia. Finally, the most unusual of all, the marsupial moles (*Notoryctes typhlos* et *N. caurinus*), are so highly specialized that they warrant a separate order. Morphologically, they resemble African golden moles (family Chrysochloridae) more than opossums or kangaroos.

Extinct representatives of all four orders have been discovered in Oligo-Miocene localities in Australia. Additionally, these faunas include species without modern equivalents. Some can be placed within established groups based on anatomical features, but other species are so distinct that entirely new orders have been proposed (see Sect. "The Curious Yalkaparidons", in Appendix, and Fig. 6.2).

We will proceed systematically in exploring Australia's ancient fauna. Later, in our "prehistoric excursion," I will highlight the most significant comparisons and differences between present-day marsupial faunas and the fossil ones found in Queensland, commonly referred to as the Riversleigh fauna, named after the principal site of discovery.

Many modern marsupials are adapted to open, sometimes arid environments, reflecting today's typical Australian conditions. Yet there are also forested regions, home to species such as koalas, some quolls, and wallabies,

Fig. 6.2 Left, *Yalkaparidon coheni* (Lower Miocene); currently, only the skull of the animal is known. Right, fossil marsupial mole, *Naraboryctes philcreaseri* (Lower Miocene), weighing about 200 g. The two animals are not shown to scale

mainly in the northern part of the continent and along the western coast, which receive seasonal monsoon rainfall.

During the Oligo-Miocene, however, Australia's climate was markedly different: tropical forests covered at least the entire northwestern region. The fauna of the period thrived in these environments, now much rarer.

Although today's koala inhabits relatively open, dry eucalyptus forests, fossil species appear to have preferred the lush forests of the Miocene. Several species are known from the Oligo-Miocene, not all contemporary. Over time, their numbers declined, likely due to continental aridification, leaving only one surviving species. Despite their greater past diversity, koalas are relatively rare in the Riversleigh fossil record.

The fossil marsupial mole (*Naraboryctes philcreaseri*) presents an interesting case. Modern species inhabit deserts, "swimming" through sand (Archer et al. 2010), looking for prey (mainly arthropods). The Miocene species (Fig. 6.2), however, lived in tropical forests. Does this imply different habits or morphology? Not necessarily. Despite anatomical differences, fossil and modern species share enough traits to belong to the same order and lead similar lifestyles. The fossil mole "swam" through loose forest soil, not unlike sand in density. In other words, it was "pre-adapted" to the desert conditions that developed later, easily adjusting to shifting sands as its ancestors navigated loose forest substrates.

Kangaroos are among Australia's most emblematic animals. The best-known species, the red kangaroo (*Osphranter rufus*), is an herbivore perfectly adapted to arid environments. Kangaroos can be considered ecological equivalents of gazelles or cattle in the Old World, with herbivorous, frugivorous, and even fungivorous diets. Some species prefer open habitats, others more closed or forested environments, and some (genus *Petrogale*) inhabit rocky, steep areas similar to ibex and chamois in mountains.

During the Oligo-Miocene, Australia's environment was more forested, and kangaroo diets were more varied. Some species possessed specialized dentition, with a rasp-shaped premolar, more or less developed, on each side of their jaws. Palaeontologists think that the diet of these animals must have been clearly omnivorous, if not carnivorous, since their teeth were perfectly shaped to cut meat foods. However, at the moment, this remains a hypothesis, although it is justified by the size and shape of their premolars.

These kangaroos form a particular subfamily, the Propleopinae (Fig. 6.3), related to the musk kangaroo rat (*Hypsiprimnodon moschatus*), and existed from the Late Oligocene (~28 My) to the end of the Pleistocene (~10,000 years ago). Early species were modest in size, but later forms, such as *Propleopus oscillans*, reached 70 kg, comparable to a modern grey kangaroo (*Macropus giganteus*).[32] Being considered a carnivorous, what prey could it hunt?

Another kangaroo subfamily specialized in browsing tall foliage. As forests shrank due to aridification, competition for leaves intensified. Some species evolved strong forelimbs with curved claws to grasp branches and feed efficiently, growing increasingly large. *Procoptodon goliah*, for example, exceeded two metres in height and could reach branches 2.5–3 m above the ground (Fig. 6.4).

In the forests, koalas were not the only arboreal mammals. A host of possums, related to modern species (the "possums" of the Australians) or belonging to fossil-only groups, occupied the forest canopy for fruits and other foods.

Bandicoots were also abundant in forests, with smaller species likely insectivorous and larger species carnivorous. Some researchers suggest that they

Fig. 6.3 Reconstruction of the Miocene Queensland propleopine kangaroo *Ekaltadeta ima*, considered omnivorous, weighing about 15 kg

[32] Despite the scientific name "*M. giganteus*", the grey kangaroo is not the largest. The red kangaroo has a more important mass: large males can exceed 80 kilos!

Fig. 6.4 Reconstruction of a female Pleistocene *Procoptodon goliah* feeding in a tree, standing over two metres tall

filled ecological roles similar to rodents, which were absent in Australia at the time. This diversity in diet likely explains their abundance.

Carnivorous marsupials, the dasyuromorphs, are present in the Riversleigh sediments but are relatively rare compared to today, their ecological role perhaps partially filled by bandicoots. Many ancient forms cannot be linked directly to modern families; only from the Pliocene (~5 Ma) do fossils identifiable with modern tribes appear.

But there is one exception; the thylacine family. Representatives of this group have been found in the Oligo-Miocene levels of Queensland and various species, belonging to different genera, have been described in the Riversleigh faunas. Generally smaller than the recent species (which became extinct during the nineteenth century; Paddle 2002), fossil thylacines indicate that the family was typical of forest environments, at least in times past.

Oligo-Miocene forests also hosted herbivorous marsupials without modern equivalents, adapted so well to the Tertiary environment that their decline and eventual extinction occurred only by the end of the Pleistocene (~10,000 years ago). Some resembled wombats but were taxonomically distinct, ranging in size from sheep to cows. The largest and most recent

Fig. 6.5 *Diprotodon optatum*, the largest known marsupial; it lived in Australia during the Pleistocene

species, *Diprotodon optatum*, which lived during the Pleistocene, reached the size of a rhinoceros (Fig. 6.5)!

Alongside these large herbivores were the palorchestids, even more bizarre in appearance. These animals had forelimbs longer than their hind limbs, all equipped with powerful claws. But the most unusual aspect was undoubtedly the short trunk they had on their snouts.[33] In short, these animals resembled a kind of "marsupial tapir" (Fig. 6.6). They had to roam the forests, searching for plant foods on the ground. Their strong claws allowed them to dig up bulbs and roots, or to manipulate other parts of plants.

All these peaceful herbivores thrived as long as Australia's climate supported relatively lush vegetation. But as aridity expanded, their populations began to decline. The Pleistocene ultimately brought an end to their existence.

Wherever herbivores are abundant, carnivores follow. Australia was home to a vast array of ferocious predators from different groups (snakes, crocodiles, thylacines, and possibly large predatory birds—although proof of their carnivory is still needed). The most fearsome of all, however, were the "marsupial lions" (Fig. 6.7).

These marsupial predators belonged to two now-extinct families. The most recent species, *Thylacoleo carnifex*, lived during the Pleistocene and was the size of a small lion. Older species ranged in size from lynx to leopard. Surprisingly, these animals descended from herbivorous ancestors. Phylogenetically, they were related to vegetarian Australian marsupials (koalas, kangaroos,

---

[33] The proboscis derives from the hypertrophic association between the upper lip and the nasal buds. Being made up of flesh and other soft tissues, it does not fossilize. However, palaeontologists are able to deduce its presence from the recoiled position of the nasal openings on the skull, as is the case with elephants and tapirs.

Fig. 6.6 Reconstruction of *Palorchestes anulus* (Miocene, Queensland; Australia). The animal was calf-sized, but more robust

Fig. 6.7 Reconstruction of *Thylacoleo crassidentatus* from the Early Pliocene (~5 My) of Queensland (Australia). This animal was leopard-sized

possums, and diprotodonts) rather than to carnivorous marsupials such as quolls.

Their ancestors had already lost the lower canines, which in most carnivorous mammals form the mandible's tusks. Marsupial lions transformed their lower incisors into bayonet-like structures, which opposed the upper front teeth and functioned like canines. Their most formidable weapons, however, were elongated premolars (two upper and two lower) acting as scissor-like blades perfect for cutting meat.

Robustly built, these predators likely hunted large herbivores of their time, such as kangaroos and diprotodonts. They dominated Australian ecosystems from the Late Oligocene through the Pleistocene and eventually disappeared, likely following the extinction of their primary prey.

Cenozoic marsupials, like their modern descendants, occupied nearly all terrestrial ecological niches. One exception was the aerial niche, filled by bats. Soon, under the canopy of the Australian forests, bats begin to flutter. Some modern families (including one, currently monotypic—i.e. composed of a single species, *Mystacina tuberculata*—and limited to New Zealand) were already present during the Oligo-Miocene. Others, however, made their appearance only in more recent times.

Bats can fly and not only hindered by the barriers constituted by the arms of the sea, as is proven by the presence of these animals on various oceanic islands (for example, the Seychelles islands, Fiji, etc.). But, the first bats colonized Australia, presumably, before its complete separation from the other southern continents. In fact, we find one of the oldest known species of bat at the site of Murgon. So, bats had already established themselves in Australia and had arrived there before the age of 40 My, although we do not know how many different waves of these animals have hit the island. But it is thanks to their presence that it was possible to establish the first relative dating of Australian Oligo-Miocene sediments, by comparing the faunas with those of other continents where similar forms lived.

Starting from the Late Oligocene, for millions of years, bats were apparently the only placental mammals sharing habitats with monotremes and marsupials. Rodents arrived only in the Pliocene, after 5 My (Long et al. 2003), coming from Asia. The group was so successful that, currently, it is the largest and most widespread in Australia.

After the rodents, and only during the Pleistocene, came humans and the dingo (*Canis lupus dingo*), which seems to be a close relative of the domestic dog. It is not yet clear whether this species was brought by humans or if, instead, it was able to reach Australia by its own means. All other terrestrial placentals (rabbits, pigs, and even dromedaries), currently present in Australia, are of recent anthropogenic import.

This short excursion to discover the Australian mammals of the past showed us the richness of the faunas of the time. It should be added that the mammals of Riversleigh were by no means alone, but shared the environment with many other creatures: insects, fish, amphibians, reptiles and, of course, birds (Archer et al. 1994). Some organisms are related to modern species, while others belong to extinct groups, such as armoured turtles, meiolaniids (real reptilian "tanks"), and giant birds like dromornithids, previously evoked.

Unfortunately, most of these magnificent Cenozoic animals no longer exist. Why did they disappear? They were part of the victims of the extinction that occurred at the end of the Pleistocene, which also affected megafauna (fauna composed of large mammals) in other parts of the world (Martin and Klein 1984). At about the same time, woolly mammoths (*Mammuthus primigenius*), mastodons (*Mammut americanum*), woolly rhinoceroses (*Coelodonta antiquitatis*), giant deer (*Megaloceros giganteus*) of northern latitudes, giant sloths (genera *Megatherium* and *Glossotherium*) and glyptodonts (genus *Glyptodon*) of South America, desappear along with other species less known to the general public.

Regarding the causes, opinions differ between those who consider that modern man (who had entered the scene several tens of thousands of years earlier) was the main responsible for the massacre and those who, on the hand, think that the extinctions are due to the climate changes that occurred at the end of the last ice age.

As always, in these cases, the causes had to be multiple and complex. Certainly, climate and environmental changes have had their weight, but, probably, other events could also have been added to make the toll of extinctions heavier.

6.3 Super-Society

In all books on life's history, the final pages are devoted to primates. However, if I end with man, it is only to introduce a new concept that would have manifested itself only in recent times. There are already several volumes that talk about primates and the history of our genealogy.[34] I would just like to point out a few main points here and, possibly, some new features that have not yet been sufficiently publicized by recent popular works.

[34] Among the various books devoted to fossil and modern primates, I recommend the following two volumes: Fleagle (2013) and Werdelin and Sanders (2010). The history of man can be traced in the work devoted to the subject and published under the direction of Coppens and Pic (2001), plus the numerous articles and volumes of French popular magazines dedicated to hominids (indicated in chronological order of publication): Collective work (2006a, b, 2007, 2012).

Primates are mammals characterized by opposable thumbs, stereoscopic vision,[35] and nails instead of claws. Their oldest remains date to the Early Eocene in Eurasia, North America, and Africa.

Present-day primates include lemurs and related species, tarsiers and monkeys. Tarsiers and monkeys belong to the haplorrhines (Haplorrhini)[36] (opposed to strepsirrhines [Strepsirrhini], which includes the remaining species), sharing dental, cranial, and reproductive traits. Monkeys and tarsiers diverged from a common ancestor distinct from lemurs.

Some years ago, the journal Nature proclaimed the discovery of the oldest known haplorrhine skeleton (Ni et al. 2013), *Archicebus achilles*. It is a tiny animal (its mass is estimated at only 20–30 g) from Chinese sediments aged between 56 and 54 My. Its dentition reveals an insectivorous diet and the postcranial anatomy suggests arboreal abilities. From the size of the orbits, the authors of the paper judge that it must have had diurnal habits (Fig. 6.8).

If the oldest haplorrhine (monkeys + tarsiers) lived in Asia, where did the first real monkeys come from? Specialists are divided between supporters of the Asian continent and those of Africa. However, it must be recognized that fossils of ancient primates are relatively rare and are generally limited to dental remains. Instead, the best characters for identifying monkeys are found on the skull. Anthropoids (the primates called "superior": apes, in the broad sense) have, behind the orbit, a complete bone wall. In other primates, such a wall is not there or is not complete (this is the case of tarsiers). To correctly diagnose an anthropoid, therefore, a skull is needed, more or less complete, and in good condition.

Some indisputable fossil apes (anthropoids) inhabited the forests that covered part of the present-day Fayum region of Egypt during the latter part of the Late Eocene (~ 37 My; Seiffert et al. 2010). However, given the diversity of the animals, it is likely that they were preceded by even older species, which lived in Africa or Asia, or on the two continents at the same time. The hunt for the complete skull of an Eocene monkey is still on!

Also in Africa, during the Oligo-Miocene, tail-bearing monkeys differentiated from hominoids (gorillas, chimpanzees, orangutans, gibbons, humans,

[35] Stereoscopic vision allows the perception of reliefs and therefore the evaluation of the exact distances from objects (an essential property for arboreal animals, such as primates). The position of the eyes, directed forward (and not to the sides) is an anatomical arrangement elaborated in vertebrates to achieve it.

[36] Among the characteristics that distinguish the two groups, there is one that must be noted: the conformation of the nose. Lemurs have a muzzle with a moist nose like those of dogs or cats (this is the condition of strepsirrhines). On the other hand, tarsiers and monkeys (haplorrhines), possess an authentic "human" nose, well separated from the upper lip (Fleagle 2013).

Fig. 6.8 *Archicebus achilles*, the oldest known primate (55 My, lower Eocene, China)

and related fossils). The latter are grouped in the group of hominoids.[37] Some of these species later migrated to Eurasia (~16 Ma).

But let us stay in Africa and take an interest in the primates of the Upper Miocene (after 10 My). During this period, certain African hominoids became bipedal[38]: the bipedal locomotion of African primates was born, and also the evolutionary line that reaches down to us.[39] During the following eras, we see a diversification of bipedal hominoids, although they retained

[37] The monkeys typical of South America belong to another group of apparently African origin, but which separated from the Old-World branch before the dichotomy described above. These primates would have reached South America during the early Oligocene.

[38] In fact, the oldest bipedal primate is not African. It is called *Oreopithecus* (*Oreopithecus bamboli*) and lived, 8 million years ago, in Tuscany (Italy). It is not an ancestor of man but, simply, a hominoid primate who had acquired the erect stature in a completely different context. Its locomotion was probably not exclusively bipedal; its forelimbs were perfectly adapted to suspension from tree branches. From what we know, the *Oreopithecus* left no descendants. However, this case shows that bipedal posture has not only developed more than once in the primate group but, probably, constitutes a trend that affects the entire hominoid group. That said, a recent work (Fuss et al. 2017) puts forward the hypothesis that the geographical area where the division between the human and great ape lineages took place should not be limited only to Africa, but should also take into account Europe.

[39] I repeat: if it is inherent in our evolutionary line, it is not to give it an exaggerated importance with regard to other organisms, but only because it allows me to introduce a final evolutionary concept which, I think, may have its weight for the future history of living beings.

the ability to climb trees. The most famous of these beings are the australopithecines. We know of various species, known in East and South Africa, and also in Chad, between 4.2 and 1.4 My. They used different habitats and resources. In this way, more than one species could live in the same environment.

The history of our evolution is not based on a linear succession of forms that, starting from a given ancestor, comes to us through various intermediate stages. It consists of a series of episodes in which various species of hominids were contemporary, some dividing the environment.

An African bipedal species (a population of australopithecines? A form still unknown?), after 3 million years, acquired a peculiar feature: the cranial part of the head became more voluminous than the facial one (the skull acquires larger dimensions than those of the face): the genus *Homo* appeared. Between 2.4 and 2 My, East Africa is inhabited by at least two species of humans, *H. habilis* and *H. rudolfensis*, clearly recognizable from their bony remains (MacLatchy et al. 2010).

Around 2 My, Africa seems to become too small for man, who leaves it to conquer the world.[40] Our planet is therefore "invaded", little by little, by these new conquerors, the members of the genus *Homo*. Some species follow one another, others are contemporary, in different places or regions. For example, about 35,000 years ago, Europe was inhabited by Cro-Magnon man (modern man, or *Homo sapiens*, the species to which we belong) and Neanderthal man (*Homo neanderthalensis*).[41] Although the bone remains of the two men have never been found at the same time in the same location, various indices show us that they would have met, at least in the Near East.

Elsewhere, the small island of Flores was inhabited by a completely different human population, which received the scientific name of *Homo floresiensis*. Its main feature was its size. In fact, this population was made up of tiny individuals: about one metre tall! The man from Flores, nicknamed the "Hobbit" because of his size, lived between 95,000 and 60,000 years ago. It is unknown outside its island environment.

---

[40] Man (or another primate capable of making tools—an *Australopithecus*?) may have left Africa a little earlier than 2 My. At the Chinese site of Longguppo, some finds from an alleged lithic industry would have been exhumed from sediments considered to be older than 2 million years (Collective work 2006a, p. 56–61). Have these finds been correctly dated? Are they really of anthropogenic origin? Who would have made them? See also Scardia et al. (2019).

[41] In this text, I use "Neanderthal man" or "Neanderthal" as synonymous.

During a certain era, therefore, at least three different human species[42] inhabited our planet. Now the whole globe is colonized by a single species, ours. How did we get to this situation? What about all the other species of *Homo*? What happened to them?

The oldest populations of anatomically modern man have been found in Africa, starting at ~ 300,000 years. They would have started to leave the continent between 120,000 and 90,000 years, heading for the Levant. They have been reported in Europe since 35,000 years ago. At the time, the continent was still inhabited by a few Neanderthals.

According to some specialists, to understand how our species formed, we need to look further afield and consider all human populations since their first exit from Africa, about 2 million years ago. All these, although considered different species, would have maintained a certain degree of interfecundity between them. A population that had just come out of Africa could have found that the reproductive isolation with respect to another human group (considered a different species and, therefore, in principle, reproductively isolated) encountered on its way was not perfect, especially in the geographical areas close to the place of speciation. In this case, some individuals of the first population could have had children with those of the second.

Therefore, "anatomically" modern man would have evolved in Africa, approximately 300,000 years. Instead, "genetically" modern man would have produced himself through "hereditary legacies" between different populations outside this continent, through a limited, but still possible, fecundity between these different groups.

This hypothesis could explain the persistence of some hereditary traits within populations of different species, and also the appearance of certain anatomical traits common in different geographical areas (Collective work (2006a) p. 62–67, Stewart and Stringer 2012). Genetic studies suggest Neanderthals contributed 1–4% of their nuclear DNA to modern Europeans and Asians (even some American populations would be involved, but none African; hybridization would have taken place outside the Black Continent; Collective work 2012).

---

42 In reality, things could be even more complex. In fact, DNA analysis taken from the remains of prehistoric humans would have detected another human group in Eurasia. This study was carried out on bone vestiges found in Denisova (southern part of Siberia), in the Altai mountains. Early results indicate that this population was genetically different from both Neanderthals and Cro Magnons. It would be another human "species", among those that populated the continent between 50,000 and 30,000 years (Collective work 2012, p. 20–25). But genetic analyses are not enough to describe a species, in the absence of valid anatomical descriptions. Especially when we do not yet know the genetic variability of Neanderthals (nor that of the presumed Denisovan man).

So, would modern man have "assimilated" the other populations present on the globe, in particular those of the Neanderthals, making them disappear little by little? In reality, genetic data does not tell us this. First, comprehensive comparative studies were carried out on a sample of prehistoric population too small to draw categorical assertions.

As for Neanderthals and Cro-Magnons, some hybridization was not only possible, but would actually occur. But the genetic results would rather be in agreement with a hybridization carried out in the regions of the Levant, at the time of the exit of modern populations from Africa and not later, during their stay in Europe. These results, taken individually, do not allow us to draw conclusions about the disappearance of the first species.

Neanderthals could have been exterminated by newcomers (modern humans) or they could have succumbed to the climate changes that followed the last ice age, or even for a different cause. It is likely that it was climatic and environmental turmoil that got the better of the last Neanderthal populations, rather than competition with Cro-Magnons.

The phenomenon is complex and the answer, without a doubt, is not simple. As in other similar events (biological crises), it is likely that the causes were multiple and that they even added together to cause lethal stress for Neanderthals (and also for other human populations different from ours).

In any case, for at least 10,000 years or so, the only human species on the planet has been ours: *H. sapiens*. We have come a long way since we entered the scene!

Since our first steps on Earth, we have left countless traces, which have become increasingly important especially after the establishment of agriculture, when man's mark on the environment became predominant in sediments and habitats.[43]

In his intra-specific interactions, man has been favoured thanks to a trump card, the articulated language (which derives from both his anatomical and

---

[43] The impact of man on the environment has grown enormously, especially during the last few decades, both in the alteration of existing natural habitats and in that of the creation of new biotopes. As an example of the second claim, just consider the accumulation of plastic waste (called PMD) in the oceans. Such waste forms islets which, on the one hand, can have harmful effects on aquatic flora or fauna, and on the other hand, host a new community of organisms. This includes various autotrophic and heterotrophic strains of bacteria, protozoa and also possible pathogens. The first studies on PMDs have shown that, despite the increase in plastic evacuated annually into the oceans, the total mass of waste seems to remain more or less constant. This means that agents capable of accelerating the natural degradation of plastic would be active. This hypothesis is supported by the presence, on the garbage islets, of some strains of bacteria capable of degrading hydrocarbons (Zettler et al. 2013). In any case, new analyses are needed to understand the real extent of this discovery. However, I cannot refrain from pointing out that representatives of the domain of bacteria are capable of thriving in all situations, whether natural or man-made (remember what was said in paragraph 2.3?).

intellectual abilities) which has allowed him to constitute complex societies with a myriad of different interactions between individuals. These reports have gone far beyond what has been said so far.

Human societies began to come into contact with each other very early in human history and had to learn to interact in an increasingly close way. The process has been gradual, but now, with the phenomenon of globalization, a human super-society has been created whose units would no longer be isolated individuals, but societies themselves. A new structure has been formed, adding the units of lower order: D + D + ... + D = E, where "D – D – ... – D" are the different human societies and "E" the super-society.

Is this phenomenon limited to humans? Maybe not. A similar phenomenon occurs in nature with the Argentine ant (*Linepithema humile*). This insect native to South America (it was described for the first time around Buenos Aires) has now been introduced all over the globe, thanks to international trade: first in North America, then in Europe, Africa, Australia and, finally, in Asia.

If in its country of origin, this ant leads a life without stories, in the conquered territories it reveals a very different character. It becomes capable of monopolizing local resources not only to the detriment of other hymenopterans species but also of other arthropods.[44]

The key to its success would be biological and behavioural. In the nests of Argentine ants, several queens can cohabit. Not only that, but these do not reproduce in flight, like most other species, but underground, sheltered from predators that could reduce their numbers.

However, the change that has taken place outside its home area is even more spectacular. Studies carried out in North America have shown that the different colonies of Argentine ants do not fight each other, but cohabit as if they were part of an even larger structure; the super-colony.

At first, it was thought that the reason for this behaviour was genetic. It was believed that ants in each importing country belonged to a small number of genetic strains (only two for Europe) and that, therefore, all related individuals exhibited pacifist behaviour despite living in different colonies.

In reality, all this is not true: the genetic variation in the colonized continent is not at all inferior to that they have in their country of origin. So how do you explain this change? The formation of non-warlike relations

[44] On this subject, I cannot refrain from mentioning a literary work by an author I appreciate very much: Italo Calvino (Calvino 1975). This story was pointed out to me by Professor Bruno Corbara of the *Université Blaise Pascal*, in Clermont-Ferrand, who I discovered to be a fervent admirer of the Italian novelist. If the reading of the text does not add anything to the scientific discourse, it at least has the merit of being pleasant to read (like all his works) and showing us the way in which this small insect was considered during its colonization of northern Italy.

and the peaceful cohabitation of different colonies could be the result of a strong selection to recognize the most frequent chemical signals that can be encountered in the conquered territories (added to the fact that several queens cohabit in the same nest; Giraud et al. 2004).

As you can see, this is not exactly what happens to humans. The human super-society has a different structure from that of the super-colonies of the Argentine ant and the types of signals used are quite distinct (articulated language, in one case, chemical and tactile signals in the other).

Are they, therefore, two very different phenomena or simply two aspects of the same phenomenon? It is too early to say. To answer this question, more precise studies are needed, especially to better define the characteristics of human super-societies and the system of open ant colonies.

Despite this, these super-colonies are very real. They allow the Argentine ant to be competitively superior to native species and other predators, such as spiders.

But the story does not end there. Another ant species, *Lasius neglectus*, began invading Europe from the Asian steppes. It too is capable of forming giant colonies with multiple queens and of eliminating competitors. The super-colony system could therefore be a feature not limited to the Argentine ant and could become (if it has not already done so) the key to the success of certain social insects against their competitors.

However, let us not jump to conclusions. Insect super-colonies have only recently been discovered and may represent a relatively recent evolutionary phenomenon. Nevertheless, they function effectively. The phenomenon of super-societies could produce significant changes within communities and plant populations and have a non-negligible impact on biodiversity. In this case, we will have the opportunity to observe the evolution of the process first-hand. As with many other natural phenomena, this is a case worth following closely.

6.4 Summary of This Chapter

The Cenozoic is the most recent era. It spans the period between the end of the Mesozoic (~66 My) and the present day. It is also the era we know best, both from paleontological and climatological perspectives. In this regard, data collected during the Quaternary glacial periods are crucial for understanding the evolution of modern climate and temperatures. These data are based on isotopic analyses from ice samples taken from polar ice caps or from the shells of foraminifera living at the time under study.

The Cenozoic is also called the "age of mammals" because mammals took the place of the now-extinct dinosaurs. During this era, mammals evolved, increasing in size compared to the Secondary period. But they were not alone: birds, direct descendants of dinosaurs, were also capable of producing large terrestrial species. Giant avian forms are known throughout the Cenozoic, and it seems that they could also have been present during the late Cretaceous.

Regarding the evolution of mammals, two examples are presented. The first leads us to the discovery of the inhabitants of Australia, where one of the two most important groups that survived the K/T crisis, that of the marsupials, has been able to occupy almost all the terrestrial ecological niches used by mammals. Only that of the bats was missing from them. Bats arrived on the continent during the Palaeogene (~ 50 My). From the same period, marsupials also began to diversify. Today, representatives of all four extant orders have been identified, along with some species that became extinct without leaving descendants.

The second example concerns primates. These arboreal animals began to diversify during the Paleogene, the first period of the Cenozoic. The earliest known monkeys, based on relatively complete remains, were found in Egyptian sediments from the late Eocene (~ 37 My). Since they were already diverse, specialists believe that even older species must have existed, though their exact regions remain uncertain. Africa is also the continent where tailed apes diverged from tailless monkeys (the group including anthropoids and humans) and where the oldest remains of modern humans (*Homo sapiens*) have been found, dating to about 300,000 years ago.

When modern humans left Africa, the world was inhabited by other human species, including Neanderthals, whom our ancestors undoubtedly encountered in Europe and/or the Near East. Despite this diversity, *Homo sapiens* is currently the only human species present on the planet, a situation that has persisted for at least 10,000 years.

Humans have built complex societies, thanks to their unique form of communication: articulated language. Such close relations between societies have given rise to a super-society, where the basic units are the societies themselves.

Another species capable of similar social complexity is a small insect: the Argentine ant. Members of different colonies, outside their original range, are able to recognize each other as part of the same intraspecific entity. These habits seem to provide various advantages over competing arthropods in exploiting environmental resources.

The formation of super-societies could therefore represent a new evolutionary phenomenon with important implications for social species capable of implementing it. This development seems very recent (about two centuries, considering the Argentine ant). However, before speculating further, more data are needed in this domain.

Appendix

Current Marsupials

Modern marsupials are divided into two major groups: Australidelphia, which includes all Australian forms (plus one particular species, *Dromiciops gliroides*, the mountain monito from Argentina and Chile), and Ameridelphia, which includes opossums and caenolestids from the Americas. These groups differ in skeletal characteristics (particularly in the ankle joint), allowing identification from fossil remains (Szalay 1994). Both had a common ancestor, though it has not yet been identified.

The Curious Yalkaparidons

The enigmatic *Yalkaparidon coheni*, including the species *Y. jonesi*, is the sole representative of the order Yalkaparidontia (Archer et al. 1988), a taxonomic group restricted to the Early and Middle Miocene (23–11 My) of Australia. These animals are characterized by unusual dentition: a central pair of powerful bayonet-shaped lower incisors, combined with small crescent-shaped jugal teeth (molars and premolars). No other marsupial displays this arrangement. It is hypothesized that these animals fed on soft-bodied invertebrates or pierced eggs with their incisors. As no postcranial skeleton has been found, reconstructing their lifestyle remains speculative, although complete skulls of both species have been discovered.

Neanderthal

Neanderthals lived between 300,000 and 28,000 years ago in Europe, the Near East and western Asia (though eastern Asian populations may have affinities with this species). Anatomically, they had a large, elongated skull (its cerebral capacity exceeds ours: 1750 cm^3 against only 1500 cm^3!), prominent supraorbital ridges, and an occipital protuberance. They were more robust and stockier than modern humans, with cylindrical chests. Neanderthals were skilled hunters (preying on animals from rabbits to mammoths), expert toolmakers (a locality in the Allier, Châtelperron, was used to baptize his most recent production) and that he buried the dead (a practice attested in various localities). They were perfectly adapted to their environment and survived for over 250,000 years (nearly as long as modern humans!). For accessible bibliographic references, see the works listed in chronological order: Collective work (2006c), (2011), (2012).

References[45]

*Agusti, J., and M. Antón. 2002. *Mammoths, sabertooths, and hominids*. New York: Columbia University Press.

Alvarenga, M.F., and E. Höfling. 2003. Systematic revision of the Phorusrhacidae (Aves: Ralliformes). *Papéis Avulsos De Zoologia* 43: 55–91.

Andors, A. 1992. Reappraisal of the Eocene groundbird *Diatryma* (Aves: Anserimorphae) in *Papers in avian paleontology honoring Pierce Brodkorb. Natural History Museum of Los Angeles County Science Series* 36: 109–125.

Angst, D., C. Lécuyer, R. Amiot, É. Buffetaut, E. Fourel, E. Martineau, S. Legendre, A. Abourachid, and A. Herrel. 2014. Isotopic and anatomical evidence of and herbivorous diet in the early Tertiary bird *Gastornis*. Implications for the structure of Paleocene terrestrial ecosystems. *Naturwissenschaften* 101: 313–322.

Archer, M., S.J. Hand, and H. Godthelp. 1988. A new order of Tertiary zalambdodont marsupials. *Science* 239: 1528–1531.

*Archer, M., S.J. Hand, and H. Godthelp. 1994. *Riversleigh: The story of animals in ancient rainforest of Inland Australia*. Sydney: Reed Natural History.

Archer, M., R.M.D. Beck, M. Gott, S.J. Hand, H. Godthelp, and K. Black. 2010. Australia's first fossil marsupial mole (Notoryctemorphia) resolves controversies about their evolution and palaeoenvironmental origins. *Proceedings of the Royal Society* 278: 1498–1506.

Beaumont, A., and P. Cassier. 1994. *Biologie animale—les Cordés, anatomie comparée des Vertébrés*, 6th ed. Paris: Dunod.

Beck, R.M.D. 2012. An ameridelphian marsupial from the early Eocene of Australia supports a complex model of Southern Hemisphere marsupial biogeography. *Naturwissenschaften* 99: 715–729.

Bellwood, D.R. 1996. The Eocene fishes of Monte Bolca: The earliest coral reef assemblage. *Coral Reef* 15: 11–19.

Bibi, F., A.B. Shabel, B.P. Kraatz, and T.A. Stidham. 2006. New fossil ratite (Aves: Palaeognathae) eggshell discoveries from the Late Miocene Baynunah formation

[45] Works preceded by an asterisk can be read, more or less easily, by people with a non-professional skilling.

of the United Arab Emirates, Arabian Peninsula. *Palaeontologica Electronica* 9 (1): 1–13.

Boles, W.E. 1997a. Hindlimb proportions and locomotion of *Emuarius gidju* (Patterson and Rich, 1987) (Aves, Casuariidae). *Memoirs of the Queensland Museum* 41 (2): 235–240.

Boles, W.E. 1997b. Fossil Songbirds (Passeriformes) from the early Eocene of Australia. *Emu* 97 (1): 43–50.

Boles, W.E. 2001. A new Emu (Dromaiinae) from the late Oligocene Etadunna formation. *Emu* 101 (4): 317–321.

Bourdon, E., and J. Cracraft. 2011. *Gastornis* is a terror bird: New insights into the evolution of the Cariamae (Aves, Neornithes). *Society of vertebrate paleontology 71st annual meeting program and abstract*, 75.

Buffetaut, É. 1997. L'oiseau géant *Gastornis*: Interprétations, reconstruction et vulgarisation des fossiles inhabituels dans la France du XIXe siècle. *Bulletin De La Société Géologique De France* 168 (6): 805–812.

Buffetaut, É. 2011. *Samrukia nessovi*, from the late Cretaceous of Kazakhstan: A large pterosaur, not a giant bird. *Annales De Paléontologie* 97: 135–138.

Buffetaut, É., and J. Le Loeuff. 2010. *Gargantuavis philoinos*: Giant bird or giant pterosaur? *Annales De Paléontologie* 96: 135–141.

*Calvino, I. 1975. *The watcher & other stories*. San Diego: Harcourt Brace Jovanovich.

*Collective work. 2006a. *La nouvelle histoire de l'homme*. Sciences et Avenir, 710 (April).

*Collective work. 2006b. *La nouvelle histoire des hommes disparus*. Sciences & Vie, H.S. 235 (June).

*Collective work. 2006c. *Neandertal enquête sur une disparition*. Dossiers de La Recherche, 24 (August–October).

*Collective work. 2007. *Sur les traces de nos ancêtres*. Dossier pour la Science 57 (October–December).

*Collective work. 2011. *Néandertal réhabilité*. Dossier d'Archéologie, 345 (May–June).

*Collective work. 2012. *L'homme de Néandertal*. Dossier pour la Science 76 (July–September).

*Coppens, Y., and P. Pic, eds. 2001. *Aux origines de l'humanité—De l'apparition de la vie à l'homme moderne*. Paris: Fayard.

Flannery, T.F., T.H. Rich, P. Vickers-Rich, T. Ziegler, E.G. Veatch, and K.M. Helgen. 2022. A review of monotreme (Monotremata) evolution. *Alcheringa* 46 (1): 3–20.

Fleagle, J.G. 2013. *Primate adaptation and evolution*, 3rd ed. London: Academic Press.

Fuss, J., N. Spassov, D.R. Begun, and M. Bömme. 2017. Potential hominin affinities of *Graecopithecus* from the late Miocene of Europe. *PLoS ONE* 12 (5): 1–23.

*Giraud, T., L. Passera, and L. Keller. 2004. L'expansion coloniale de la fourmi d'Argentine. *La Recherche* 372: 52–54.

Godthelp, H., S. Wroe, and M. Archer. 1999. A new marsupial from the early eocene tingamarra local fauna of Murgon, southeastern Queensland: A prototypical Australian marsupial? *Journal of Mammalian Evolution* 6 (3): 289–313.

Guérin, C., and M. Patou-Mathis, eds. 1996. *Les grands mammifères plio-pléistocènes d'Europe*. Paris: Masson.

Hansen, J.E., and M. Sato. 2012. Paleoclimate implications for human-made climate change. In *Climate changes*, ed. A. Berger, F. Mesinger, and D. Sijacki, 21–47. Vienna: Springer.

Janis, C.M., K.M. Scott, and L.L. Jacobs, eds. 2005. *Evolution of tertiary mammals of North America—Volume 1: Terrestrial carnivores, ungulates, and ungulatelike mammals*, 2nd ed. Cambridge: Cambridge University Press.

Janis, C.M., Gunnell, G.F., and M.D. Uhen, eds. 2018. *Evolution of Tertiary Mammals of North America - Volume 2: Small Mammals, Xenarthrans, and Marine Mammals*, 2nd ed. Cambridge University Press.

Kemp, T.S. 2005. *The origin and evolution of mammals*. Oxford University Press.

*Long, J.A., M. Archer, T. Flannery, and S.J. Hand. 2003. *Prehistoric mammals of Australia and New Guinea—One hundred million years of evolution*. Baltimore and London: Johns Hopkins University Press.

Lozupone, C.A., J.I. Stombaugh, J.I. Gordon, J.K. Jansson, and R. Knight. 2012. Diversity, stability and resilience of the human gut microbiota. *Nature* 489: 220–230.

MacLatchy, L.M., J. Desilva, W.J. Sanders, and B. Wood. 2010. Hominini. In *Cenozoic mammals of Africa*, ed. L. Werdelin and W. J. Sanders, 471–540. Berkeley: University of California Press.

Martin, P.S., and R.G. Klein. 1984. *Quaternary extinctions—a prehistoric revolution*, 892. Tucson: The University of Arizona Press.

*Menkhorst, P., and F. Knight. 2010. *Field guide to the mammals of Australia*, 3rd ed. Melbourne: Oxford University Press.

Mourer-Chauviré, C. 1999. Les relations entre les avifaunes du Tertiaire inférieur d'Europe et d'Amérique du Sud. *Bulletin De La Société Géologique De France* 170 (1): 85–90.

Mourer-Chauviré, C., B. Senut, M. Pickford, and P. Mein. 1996. Le plus ancien représentant du genre *Struthio* (Aves, Struthionidae), *Struthio coppensi* n. sp., du Miocène inférieur de Namibie. *Comptes Rendus De L'académie des Sciences* 322 (4): 325–332.

Naish, D., G. Dyke, A. Cau, F. Escuillié, and P. Godefroit. 2012. A gigantic bird from the upper cretaceous of Central Asia. *Biology Letters* 8 (1): 1–4.

Nguyen, J.M.T., W.E. Boles, and S.J. Hand. 2010. New material of *Barawertornis tedfordi*, a Dromornithid bird from the Oligo-Miocene of Australia, and its phylogenetic implications. *Records of the Australian Museum* 62: 45–60.

Ni, X., D.L. Gebo, M. Dagosto, J. Meng, P. Tafforeau, J.J. Flynn, and K.C. Beard. 2013. The oldest known primate skeleton and early haplorhine evolution. *Nature* 498: 60–64.

Nilsson, M.A., U. Arnason, P.B. Spencer, and A. Janke. 2004. Marsupial relationships and a timeline for marsupial radiation in South Gondwana. *Gene* 340: 189–196.

Nilsson, M.A., G. Churakov, M. Sommer, N. Van Tran, A. Zemann, J. Brosius, and J. Schmitz. 2010. Tracking Marsupial evolution using archaic genomic retroposon insertions. *PloS Biology* 8 (7): 1–7.

O'Leary, M.A., J.I. Bloch, J.J. Flynn, T.J. Gaudin, A. Giallombardo, N.P. Giannini, S.L. Goldberg, B.P. Kraatz, Z.-X. Luo, J. Meng, X. Ni, M.J. Novacek, F.A. Perini, Z.S. Randall, G.W. Rougier, E.J. Sargis, M.T. Silcox, N.B. Simmons, M. Spaulding, P.M. Velazco, M. Weksler, J.R. Wible, and A.L. Cirranello. 2013. The placental mammal ancestor and the post-K/Pg radiation of placentals. *Science* 339: 662–667.

*Paddle, R. 2002. *The last tasmanian tiger—The history and extinction of the thylacine*. Cambridge University Press.

Piton, L.É. 1940. Paléontologie du gisement éocène de Menat (Puy-de-Dôme) (Flore et Faune). Mémoires de la Société d'Histoire naturelle d'Auvergne. *Clermont-Ferrand* 1: 1–303.

Prothero, D.R. 2006. *After the dinosaurs: The age of mammals*. Bloomington (Indiana): Indiana University Press.

Rössner, G.E., and K. Heissing, eds. 1999. *The Miocene land mammals of Europe*. Monaco: Pfeil-Verlag.

Ruddiman, W.F. 2003. The anthropogenic greenhouse era began thousands of years ago. *Climatic Change* 61: 261–293.

*Savage, R.G.J., and M.R. Long. 1986. *Mammals evolution—An illustrated guide*. London: British Museum of Natural History.

Scardia, G., F. Parenti, D.P. Miggins, A. Gerdes, A.G.M. Araujo, and W.A. Neves. 2019. Chronologic constraints on hominin dispersal outside Africa since 2.48 Ma from the Zarqa Valley Jordan. *Quaternary Science Reviews* 219: 1–19.

Seiffert, E.R., E.L. Simons, J.G. Fleagle, and M. Godinot. 2010. Paleogene antrhopoids. In *Cenozoic mammals of Africa*, ed. L. Werdelin and W. J. Sanders, 369–391. Berkeley: University of California Press.

*Selosse, M.A. 2017. *Jamais seul—Ces microbes qui construisent les plantes, les animaux et les civilisations*. Arles: Actes Sud.

Senut, B., and M. Pickford. 1995. Fossil eggs and Cenozoic continental biostratigraphy of Namibia. *Palaeontologia Africana* 32: 33–37.

Stewart, J.R., and C.B. Stringer. 2012. Human evolution out of Africa: The role of Refugia and climate change. *Science* 335: 1317–1321.

Szalay, F.S. 1994. *Evolutionary history of the marsupials and an analysis of osteological characters*. Cambridge University Press.

Taylor, T.N., E.L. Taylor, and M. Krings. 2009. *Paleobotany: The biology and evolution of fossil plants*, 2a ed. Amsterdam: Academic Press.

Tennyson, A.J.D., T.H. Worthy, C.M. Jones, R.P. Scofield, and S.J. Hand. 2010. Moa's ark: Miocene fossils reveal the great antiquity of Moa (Aves: Dinornithiformes) in Zealandia. *Records of the Western Australian Museum* 62 (1): 105–114.

*Turner, A., and M. Antón. 2000. *The big cats and their fossil relatives*. New York: Columbia University Press.

*Turner, A., and M. Antón. 2004. *Evolving Eden—An illustrated guide to the evolution of the African large-mammal fauna*. New York: Columbia University Press.

Vicker-Rich, P.V. 1979. The Dromornithidae, an extinct family of large ground birds endemic to Australia. *Bureau of Natural Resources, Geology and Geophysics* 184: 1–196.

*Von Koenigswald, W., and G. Storch. (eds). 1998. *Messel: la mémoire de la nature* (traduction de J.-L. Schlegel). Seuil, Paris.

*Wang, X., R.H. Tedford, and M. Antón. 2008. *Dogs: Their fossil relatives and evolutionary history*. New York: Columbia University Press.

Wang, X., L.J. Flynn, and M. Fortelius, eds. 2013. *Fossil mammals of Asia–Neogene biostratigraphy and chronology*. New York: Columbia University Press.

Werdelin, L., and W.J. Sanders, eds. 2010. *Cenozoic mammals of Africa*. Berkeley: University of California Press.

Wilf, P., N.R. Cúneo, I.H. Escapa, D. Pol, and M.O. Woodburne. 2013. Splendid and seldom isolated: The paleobiogeography of Patagonia. *Annual Review of Earth and Planets Sciences* 41: 561–603.

*Wilson, K.-J. 2004. *Flight of the Huia—Ecology and conservation of New Zealand's frogs, reptiles*. Birds and Mammals: Canterbury University Press.

Witmer, L., and K.D. Rose. 1991. Biomechanics of the jaw apparatus of the gigantic Eocene bird *Diatryma*; implications for diet and mode of life. *Paleobiology* 17 (2): 95–120.

*Wroe, S. 1999. The bird from hell? *Nature Australia* 26: 58–64.

Yuan, C.-X., Q. Ji, Q.-J. Meng, A.R. Tabrum, and Z.-X. Luo. 2013. Earliest evolution of multituberculate mammals revealed by a new Jurassic fossil. *Science* 341: 779–783.

Zachos, J., M. Pagani, L. Sloan, E. Thomas, and K. Billups. 2001. Trends, rhythms, and aberrations in global climate 65 Ma to present. *Science* 292: 686–693.

Zettler, E.R., T.J. Mincer, and L.A. Amaral-Zettler. 2013. Life in the plastisphere: Microbial communities on plastic marine debris. *Environmental Science and Technology* 47 (13): 7137–7146.

Conclusions

We have reached the end of our journey. Throughout this story, which summarizes more than 3.5 billion years of Life on Earth, I have attempted to present the evolutionary history of living beings as clearly as possible, from their first appearance to the present day. In my opinion, it is the most exciting way to understand both how current biodiversity developed and to appreciate it correctly.

Of course, given the amount of information accumulated in the field of palaeontology and evolution (especially in recent years), I could only provide a brief outline of this history. My main goal, however, was to spark curiosity about the subject and to provide some bibliographical references to allow the reader to continue exploring this domain according to their interests.

Finally, what conclusions can be drawn from this account and the data collected so far?

First of all, it is necessary to reflect on the methods of investigation of the life sciences, the domain concerned which research on present and past biodiversity. We find that many research results are difficult to interpret and always require extensive verification before they are considered reliable. In fact, each answer often raises new questions which, in turn, open new horizons. Sometimes, these questions cast doubt on certainties that initially seemed firmly established.

All this may seem disconcerting to the general public, but it is the "routine" of science. Science does not provide miraculous solutions, but follows a precise methodology that, despite seeming disappointing results, has been repeatedly tested. Every hypothesis and every result must be examined through this lens and verified by additional analyses until they find their place

A. M. F. Valli, *The Three Domains of Life*, Copernicus Books,
https://doi.org/10.1007/978-3-032-14802-5

within the coherence of formulated theories. Although many interpretations may initially appear fleeting and changeable, after verification and additional results, they can acquire the solidity initially lacked. Or, they can be replaced for more relevant explanations.

Life sciences involve a large number of variables and are extremely complex. It is normal to proceed by trial and error. However, if we compare what we knew only a few decades ago with current knowledge, we realize that enormous progress has been made. Despite overturned certainties and questions that still await an answer, the results are tangible. And all this has been made possible by the effort of thousands of researchers who work diligently and employ rigorous scientific method.

Like all other scientific disciplines, the life sciences advance in cycles composed of two equally important phases: initially, the accumulation of raw data (whether measurements on a current population or the excavation and description of fossil remains of a fossiliferous locality), followed by the synthesis of the results produced. This step can lead to the integration of data into a known theory, to the modification of that theory, or even to the development of entirely new hypotheses. The progress of recent techniques allows us assemble amounts of data (as evidenced by the huge number of scientific publications every year!). Some data gradually find their place in existing theories, while others must wait longer. But in the end, all data will be "organized" into a coherent set of ideas.

As shown in previous chapters, interdisciplinary studies (i.e., the use of different disciplines to achieve a goal) have become fundamental to scientific progress. Reconstructing the exact sequence of past events is no exception. Increasingly, palaeontologists and geologists require the assistance of specialists from other scientific fields to carry out their daily work, and they also need knowledge in disciplines other than their own, including statistics, physics, chemistry, medicine, and others.

In fact, it is no longer simply a matter of recognizing and describing the remains of extinct organisms. One must also understand their ways of life and interactions. Certain trace fossils require specialized methodologies to be revealed (especially those involving microbes or specific groups of organisms) and the technological methods to analyse them require the intervention of true experts.

Understanding the habitat in which organisms have lived requires interpreting data related to the sedimentary environment in which the remains were preserved. It is also necessary to recognize and reconstruct the chemical transformations to which they have been subjected, and, in general, a statistical treatment of the data is required to correctly interpret the results.

These steps are necessary to provide at least partial answers to fundamental questions concerning evolutionary process.

Moreover, it is precisely the comparison between ideas from different disciplines that not only makes these analyses possible but often accelerates the synthesis of accumulated data.

To conclude, I want to emphasize one last point. If we critically analyse the details presented in the previous chapters, we realize that there is no strictly rational logic in the evolution of living beings.

Of course, anatomically more complex organisms appear after the simplest ones. But if we look closely at who are the real "rulers" of the planet in current ecosystems, we discover that, apart from a few animals elevated by our anthropocentrism bias (humans, first of all) the most ubiquitous and abundant beings are indisputably representatives of the oldest groups, especially bacteria. And this has been true since the beginning of Life on Earth!

These creatures did not go extinct when more complex organisms appeared. On the contrary, they continued to evolve and adapt alongside the newcomers[1].

If we imagine being able to rewind the film that depicts the history of Life to a particular moment and then let the film of evolution resume, the organisms that we would encounter from that point could be very different from those we have seen throughout this account (Gould 2000). The evolution of living beings tells a unique story and, in its details, does not repeat itself!

Based on current knowledge, we cannot predict the evolution of biodiversity, especially over long periods. Nevertheless, the study and description of the past help us understand evolutionary "major trends", particularly when considering narrow time intervals. In addition, they advance our understanding of nature and the place we occupy within it, allowing us to live more harmoniously with one another and with the other organisms that inhabit the planet.

[1] In this regard, it is useful to remember that modern bacteria are not primitive beings at all. Like all other species of plants and animals, they are the product of more than 3 billion years of evolution. And when we are gone, they, on the other hand, will still be there.

Appendix

International Stratigraphic Chart

Eon	Era	Period	Epoch	Numerical age (Ma)
Phanerozoic	Cenozoic	Quater-nary	Holocene	Present – 0,0117
			Pleistocene	2,6
		Neogene	Pliocene	5,3
			Miocene	23
		Paleogene	Oligocene	40
			Eocene	56
			Paleocene	66
	Mesozoic	Cretaceous	Upper	101
			Lower	145
		Jurassic	Upper	164
			Middle	174
			Lower	201
		Triassic	Upper	237
			Middle	247
			Lower	252

Eon	Era	Period	Epoch	Numerical age (Ma)
Fanerozoic	Paleozoic	Permian	Lopingian	252 – 259
			Guadalupian	273
			Cisuralian	299
		Carboniferous – Pennsyl-vanian	Upper	307
			Middle	315
			Lower	323
		Carboniferous – Missis-sippian	Upper	331
			Middle	347
			Lower	359
		Devonian	Upper	383
			Middle	393
			Lower	419
		Silurian	Pridoli	423
			Ludlow	427
			Wenlock	433
			Llandovery	444
		Ordovician	Upper	458
			Middle	470
			Lower	485
		Cambrian	Furongian	497
			Miaolingian	509
			Series 2	521
			Terre-neuvian	541

Eon		Era	Period	Numerical age (Ma)
Precambrian	Proterozoic	Neo-protero-zoic	Ediacarian	541 – 635
			Cryogenian	720
			Tonian	1000
		Meso-protero-zoic	Stenian	1200
			Ectasian	1400
			Calymmian	1600
		Paleo-protero-zoic	Statherian	1800
			Orosirian	2050
			Rhyacian	2300
			Siderian	2500
	Archean	Neo-archean		2800
		Meso-archean		3200
		Paleo-archean		3600
		Eoarchan		4000
	Hadean			~4600

Geological ages on planet earth (data from the International commission of stratigraphy site)

A. M. F. Valli, *The Three Domains of Life*, Copernicus Books,
https://doi.org/10.1007/978-3-032-14802-5

Reference[2]

*Gould, S. J. 2000. *Wonderful life—The burgess shale and the nature of history*. New York: Vintage.

[2] Works preceded by an asterisk can be read, more or less easily, by people with a non-professional skilling.

A. M. F. Valli, *The Three Domains of Life*, Copernicus Books,
https://doi.org/10.1007/978-3-032-14802-5